SNIPER: ONE ON ONE

Adrian Gilbert

SNIPER:
ONE ON ONE

The World of Combat Sniping

SIDGWICK & JACKSON
LONDON

First published 1994 by Sidgwick & Jackson Limited

a division of Pan Macmillan Publishers Limited
Cavaye Place London SW10 9PG
and Basingstoke

ISBN 0 283 06165 0

PRINTED IN THE U.S.A.

CONTENTS

CONTENTS

ACKNOWLEDGEMENTS

Most books rely on the work of others beside the author, and this volume is no exception. The Imperial War Museum is a treasure trove of information, and I extend my appreciation to the keepers and staff of the Departments of Printed Books, Documents, Sound Records and Photographs. The National Army Museum was also useful. Besides its magnificent collection of small arms, the Ministry of Defence Pattern Room in Nottingham is equipped with a first-rate library. I must thank the custodian H. J. Woodend and the librarian Howard Mitchell, whose generous and well-informed help was invaluable.

In the practical sphere, I received expert guidance at the School of Infantry, Warminster. Firstly, I must acknowledge the role of Major R. G. Williams in opening doors to me. Lieutenant-Colonel 'Tug' Wilson MBE, curator of the Weapons Museum (along with Major John Old-field), was most helpful in suggesting lines of enquiry. Particular thanks go to the instructing staff (and students) of the Snipers' Division of the Tactics and Small Arms Wing. They patiently answered my questions and provided me with a chance to gain first-hand knowledge of the practice of sniping and the training of snipers.

The Royal Marines were equally helpful. Lieutenant-Colonel N. C. Thompson MBE guided me towards The Commando Training Centre, Lympstone, where I was fortunate in being assisted by the Infantry Support Wing, who carefully explained the Royal Marine approach to sniping.

In the United States, I was most grateful for the guidance given by Major Alister Reed, the military attaché at the British Embassy in Washington, and to Lieutenant-Colonel David Black, the British Liaison Officer at the US Army Infantry Center and School, Fort Benning, Georgia, Also at Fort Benning, my thanks extend to Dr Charles White, the Chief of Military History. The US Army Military History Institute at Carlisle, Pennsylvania, provided me with the fullest assistance, with special acknowledgement to John J. Slonaker, the Chief, Historical Reference Branch.

ACKNOWLEDGEMENTS

The United States Marine Corps Museum at the Navy Yard, Washington, DC, was an essential port of call, especially the Reference and Research Section, which is particularly well stocked with material on Marine Corps sniping.

From Canada I received generous assistance from firearms expert Warren Wheatfield.

At Sidgwick & Jackson, I must thank William Armstrong, for suggesting the idea for this book, and Helen Gummer, my editor.

I am indebted to Mary Gilbert-Brown and Sally Payne for their advice and help with the manuscript; and for his unwavering forbearance during school holidays, my thanks are due to Louis Gilbert.

FIGURES IN THE TEXT

INTRODUCTION

The snipe is a small game bird similar to the woodcock. In India, during the late eighteenth century, British Army officers popularized the hunting of the snipe, its small size and agility making it a difficult and rewarding target. The term 'sniper' soon came to describe a skilful hunter, well versed in the arts of stalking, and a first-rate shot. Naturally, the expression entered into the military sphere; the earliest recorded instance of its use is found in a letter from India in 1773: 'The soldiers . . . put their hats on the parapet for the enemy to shoot at, and humorously called it sniping.'

For most of the nineteenth century, the words 'sniper' and 'sniping' remained confined to the British areas of the English-speaking world; elsewhere, accurate rifle fire at selected individuals was merely called 'sharpshooting'. It was not until the twentieth century that sniping came to describe a specific military activity, separate from sharpshooting. Although the distinction remains one of degree, a sniper in the British or US armed forces is considered a specialist, whose prime function is to kill selected high-value targets at long range using superior skill and armament. A sharpshooter, by contrast, is a rifleman (proficient or otherwise) who acts in an opportunist manner, taking shots at the enemy when the chance arises.

The value of accurate shooting is universally accepted, and yet sniping has had an uncertain history. Criticism of the sniper has come from two sources. The first originates from within the armed forces, where a section of opinion has questioned the whole value of sniping. These detractors argue that time and money would be better spent in improving overall shooting standards, or should be funnelled to other important sectors of military activity. Certainly, resources are always scarce: every specialist arm could have more and better equipment, and consequently the sniper's position must be repeatedly justified.

In the pages that follow, it is hoped that the sniper's claim to be a cost-effective element within the infantry battalion will be demonstrated. However, it is important to note that, due to their often precarious position within the military establishment, snipers and their advocates tend to exaggerate the value of their discipline. Snipers do not win battles alone, but they are a useful battlefield tool for an infantry commander with the knowledge and imagination to use them effectively.

The other voice of criticism directed at sniping is a moral one. Whereas the economic efficiency argument is serious in nature – demanding a reply – the moral objection is spurious; yet it seems to have had a significant effect on the thinking of armed forces in the Western world. The basic premise of the moralist complaint is that it is somehow 'unfair' to sneak up on a man and kill him – as unacceptable as homicide in the civilian world. Yet according to Clausewitz, the definition of war is 'an act of violence intended to compel our opponent to fulfil our will'. Violence is inherently brutal, and whether injury or death is occasioned by an aimed bullet or some other agency, such as a shell splinter or high explosive, it is largely irrelevant to the victim.

Paradoxically, it can be argued that sniping is one of the more 'humane' methods of conducting warfare. The sniper rifle is the most discriminating of weapons: the sniper selects only key enemy personnel, those who are most responsible for the prosecution of military operations in the front line. Other weapon systems – such as artillery and especially military aircraft – are relatively indiscriminate killers. Mistakes are frequent and their cost devastating in lives lost.

The sniper is a highly trained individual armed with a precision weapon; both factors are equally important for the success of his mission. This book aims to describe the interaction of man and rifle, to trace the historical development of sniping from its roots in the wars of the American Revolution to the present day, and to explain how the modern sniper's training is applied to operations in the field.

A NOTE ON MEASUREMENTS

The armed forces of Britain and the United States have adopted the metric system of measurement, alongside their other NATO allies, although much of the shooting fraternity continues to use the imperial system. At present both metric and imperial exist side by side. Consequently, I have left measurements in the most commonly used form. In Chapter 15 (Rifles) respective imperial-metric measurements have been provided. For example: a sniper armed with an M40A1 rifle, weighing 14.5lb (6.57kg), is able to fire a 7.62mm (.308in) M118 bullet weighing 173 grains (11.21 grams), at a muzzle velocity of 2550fps (770m/s) to hit a target at 1000 yards (914 metres).

A guide to converting measurements is given as follows:

to convert grains to grams, multiply by 0.0648; to convert grams to grains, multiply by 15.43;

to convert pounds (lbs) to kilograms, multiply by 0.4536; to convert kilograms to pounds, multiply by 2.205;

to convert inches to millimetres, multiply by 25.4; to convert millimetres to inches, multiply by 0.0394;

to convert yards to metres, multiply by 0.9144; to convert metres to yards, multiply by 1.094;

to convert feet-per-second (fps) to metres-per-second (m/s), multiply by 0.3048; to convert metres-per-second to feet-per-second, multiply by 3.281.

PART ONE

THE HISTORICAL BACKGROUND

'Now people have been shot by platoons and in corps,
the individual will be popped at or sniped, as they call it,
from time to time.'

Letters of G. Selwyn, 1782 (*OED*, 2nd Edition, p. 856)

CHAPTER ONE

RIFLEMEN AND SKIRMISHERS

The origins of sniping date back to the eighteenth century when rifles were first used as military weapons. The concept of rifling – the cutting of spiral grooves into the inside of the barrel – in order to impart spin to a projectile was nothing new, gunsmiths being aware of this technique as early as 1500. Marksmen soon discovered the advantages of rifling over conventional smooth-bore guns: a spinning bullet was far more stable in flight and thus more accurate, and because the bullet fitted tightly into the barrel the explosive gases did not escape around the sides, causing a loss of power. During the sixteenth and seventeenth centuries rifles became popular for target shooting and hunting.

The rifle, however, had a significant problem which prevented it from being adopted as a standard infantry weapon. Whereas the smooth-bore musket could pour out fire at the rate of three to four rounds a minute in a skilled soldier's hands, the rifle took between two and three times as long to load because of the need to ram the tight-fitting ball down the length of the barrel so that it would 'grip' the rifling when fired. The introduction of a greased patch wrapped round the ball made the process faster, but the difficulties of muzzle loading were never satisfactorily overcome until the arrival of the Minié bullet in the mid-nineteenth century. Loading was further hampered by the heavy fouling of the barrel caused by the use of coarse gunpowder.

As a consequence, armies were equipped with smooth-barrelled muskets with a bore size larger than the ball; what was lost in accuracy and power was gained in a greater rate of fire and

3

ease of operation. Rifles, where they existed as a military arm, were confined to a few sharpshooting units.

Central Europe was the home of the rifle, and at the great fairs held in the trading centres of Magdeburg, Ratisbon, Basle and Prague, highly organized shooting competitions were keenly contested with substantial prizes for the best marksmen. Outside target shooting, the rifle was used by huntsmen for whom accuracy was vital in bringing down big game, such as deer, wild boar and chamois. The European rifle was even given the name 'jaeger', after the huntsmen who used it successfully in the vast estates they patrolled. Jaeger rifles were relatively short-barrelled, with large bores of between .67 and .75 of an inch. The barrel was usually octagonal with an open rear sight and a blade front sight. Some weapons were fitted with double triggers to aid accuracy.

During the eighteenth century there was a steady emigration of Germans and Swiss to the New World. Many settled in William Penn's colony (Pennsylvania) where they found greater religious freedom and economic opportunity. They took with them their rifles, and their traditional rifle-shooting competitions. In the wilderness regions – as the settlers pushed the frontier westwards – the rifle took on a new importance as a tool for survival. Expertise in the use of firearms, especially long-range shooting, became an integral part of any youth's education on the frontier.

In America, the rifle was modified for better efficiency in its new environment. As a means to save scarce and expensive powder and shot, calibres were reduced to around .40 and .45 of an inch. The barrel was also lengthened, which improved accuracy and velocity, and ensured that all the powder was burnt in the barrel before the ball left the rifle. Described by firearms author Harold L. Peterson as 'America's first original contribution to the development of firearms',[1] this long, elegant weapon was an improvement on its European predecessor. Despite its Pennsylvanian origins, the long rifle became better known as the Kentucky rifle, after the exploits of riflemen from that state at the battle of New Orleans in 1815.

Rifles had been used in the wars against the French, but the sharpshooting rifleman first came to prominence during the American Revolution (1775–83). From the outset, the American Continental Congress appreciated the potential of rifle-armed backwoodsmen to fight the British, and on 14 June 1775 the Congress authorized the raising of ten independent companies. Dressed in fringed buckskin shirts, and armed with tomahawks and scalping knives alongside their rifles, the backwoodsmen were a colourful sight. One company, under the command of Captain Crescap, gave the people of Lancaster, Pennsylvania, an exhibition of their prowess with the rifle. The event was written up in a local newspaper:

> These men have been bred in the woods to hardships and dangers from their infancy. With their rifles in their hands they assume a kind of omnipotence over their enemies. Two brothers in the company took a piece of board five inches by seven inches with a bit of paper the size of a dollar nailed in the center, and while one held the board upright gripped between his knees, the other, at 60 yards, without any kind of rest, shot eight balls through it successively and spared his brother's thighs.[2]

Such a feat of arms – and high degree of fraternal confidence – may have been exceptional, but many contemporary observers noted the shooting skills of the American rifleman. One of the most telling came from a knowledgeable opponent, Major George Hanger, a British officer and firearms expert. Hanger considered that given the right conditions (no wind, good light), a marksman could hit a man in the head at a distance of two hundred yards and gain a full body shot at three hundred. After the war Hanger wrote of how he came under fire from an American rifle sharpshooter, an uncomfortable experience which he described in some detail:

> Colonel, now General Tarleton, and myself, were standing a few yards out of a wood, observing the situation of a part of the enemy which we intended to attack. There was a rivulet in the enemy's front, and a mill on it, to which we stood directly with our horses' heads fronting, observing their motions. It was

absolutely a plain field between us and the mill; not so much as a single bush on it. Our orderly-bugler stood behind us about three yards, but with his horse's side to our horses' tails.

A rifleman passed over the mill-dam, evidently observing two officers, and laid himself down on his belly; for in such positions they always lie, to take a good shot at a long distance. He took a deliberate and cool shot at my friend, at me, and at the bugle-horn man. Now observe how well this man shot. It was in the month of August, and not a breath of wind was stirring. Colonel Tarleton's horse and mine, I am certain, were not anything like two feet apart; for we were in close consultation, how we should attack with our troops which laid 300 yards in the wood, and could not be perceived by the enemy. A rifle-ball passed between him and me; looking directly at the mill I evidently observed the flash of the powder. I directly said to my friend, 'I think we had better move, or we shall shortly have two or three of these gentlemen amusing themselves at our expense.' The words were hardly out of my mouth when the bugle-horn man behind me, and directly central, jumped off his horse and said, 'Sir, my horse is shot.' The horse staggered, fell down and died . . . Now speaking of this rifleman's shooting, nothing could be better . . . I have passed several times over this ground and ever observed it with the greatest attention; and I can positively assert that the distance he fired from at us was full 400 yards.[3]

Riflemen deliberately aiming and firing at conspicuous and important targets, such as groups of mounted officers, was a common occurrence during the war for American independence. In many ways this can be regarded as the beginning of sniping. A successful instance of this tactic occurred at the battle of Saratoga (1777) when Daniel Morgan's company of riflemen were harassing British forces. During the battle one of Morgan's men, a Pennsylvanian hunter called Tim Murphy, reputedly brought down General Simon Fraser – considered to be one of the more able British commanders – at a range of three hundred yards.

The backwoodsmen's British opponents were armed with the

Brown Bess smooth-bore musket, of which Major Hanger wrote disparagingly: 'A soldier's musket, if not exceedingly ill-bored (as many are), will strike the figure of a man at 80 yards: it may even at 100, but a soldier must be very unfortunate indeed who shall be wounded by a common musket at 150 yards, provided the antagonist aims at him; as to firing at a man at 200 yards, with a common musket, you may as well just fire at the moon.'[4]

Despite the tactical advantages of their rifles, the American backwoodsmen were only effective on the battlefield when operating in a skirmishing role. The Kentucky rifles were slow to reload and lacked a bayonet, rendering them unsuitable in the line-to-line battles which, in the end, decided the outcome of the war. In addition, the riflemen were too free-spirited to fight effectively in closed ranks; their speciality was harassing the enemy from afar.

On the receiving end of the Kentucky rifle, the British were impressed by the Americans' ability; after suffering in the early fighting around Boston, General Lord Howe reported back to London of the 'terrible guns of the rebels'. The British response was to look around for riflemen of their own. They found them in the small German states which were already supplying mercenary troops for the war in the colonies.

The German princes hired out rifle-armed jaegers to the British (at double the price of regular troops) but they were not particularly successful. Their short-barrelled rifles were inferior to the American long rifles, and some were so poor they were found to be unusable. More significantly, the troops were no match for their opponents; for the most part they were mere peasants, impressed into military service by their German masters for quick gain. With a few exceptions they were jaeger in name only.[5]

Another, more potentially productive, attempt to overcome the threat of the American rifle came from a British source. Captain Patrick Ferguson, an officer in the 70th Foot, developed his own breech-loading rifle in the 1770s. Based on the French Chaumette system, Ferguson greatly improved the speed of the screw-thread breech mechanism. The rifle was both accurate and fast to fire. During trials in 1776 Ferguson was able to loose off between four and seven shots per minute – a rate which far exceeded that of the smooth-bore musket.[6]

Suitably impressed by the rifle, the British authorities allowed Ferguson to recruit a corps of skirmishers comprising a hundred men. Ferguson led his troops into action at the battle of Brandywine Creek, but there he was badly wounded, and for reasons which are not fully clear, the rifle corps was disbanded and the breech-loading rifles withdrawn from service. Ferguson slowly regained his health, only to be killed by an American rifle bullet at the battle of King's Mountain in 1780. After Ferguson's death, the breech-loading rifle temporarily disappeared from military use. Ironically, Ferguson had proved to be a less than effective sniper on the battlefield. Before being wounded at Brandywine Creek, he had had the chance to kill General George Washington. But when Washington turned and rode away Ferguson could not bring himself to shoot the man in the back. If he had been more ruthless, the history of the United States might have been very different.

As a result of the experience gained in North America, the British Army took up the rifle again during the Napoleonic Wars (1792–1815). The first unit to be equipped with rifles was the 5th Battalion of the 60th Regiment – a jaeger unit formed mainly from Germans in British service – followed by the 95th Rifles, famous for fighting the French in the Iberian Peninsula.

In order to find a suitable rifle for these sharpshooting units, extensive trials were held in 1800 by the Royal Ordnance at Woolwich, and which included models from Britain, America and Continental Europe. The winner of the competition was the rifle submitted by the English gunsmith, Ezekiel Baker. The Baker rifle was not the most accurate of the trials, but was considered to be the most suitable for military use. Well-made and durable, it fired two sizes of ball, the larger for conventional rifle shooting, the smaller operating in the manner of a smooth-bore (with the same higher rate of fire) for close-in fighting. Effectively an improved form of the then current jaeger rifle, the Baker was short-barrelled with a large calibre of between .615 and .70 of an inch, and was fitted with a long sword-bayonet which enabled the rifleman to operate as an ordinary infantryman when necessary. Assessing the accuracy of his rifle, Baker wrote: 'I have found 200 yards the greatest range I could fire at with any certainty. I have fired very well at over 300 when the wind was very calm. At 400 and 500, I have sometimes struck the object, though I have found it to vary much.'[7]

One of the first sharpshooting exploits involving a Baker rifle took place during the British retreat to Corunna in 1808. In a rearguard action against the French, Rifleman Tom Plunkett of the 95th singled out General Colbert at the bridge of Cacabelos and killed him with a single shot. At the siege of Badajos, George Simmons wrote that a group of riflemen – 'forty as prime fellows as ever pulled the trigger' – silenced a battery of French guns.[8]

During the six years of war in the Peninsula the 95th Rifles earned a reputation as the premier unit in the British Army. Kitted out in green and black – an early form of camouflage clothing – they fought as élite skirmishers, employing cover as well as their superior firearms to good effect. In battle, their first task was to silence any enemy skirmishers, prior to shooting officers or gun crews in the open. In the British Army it is unusual for soldiers of one regiment to praise another, but Lieutenant Blakiston of the 43rd Light Infantry was moved to write:

> Certainly I never saw such skirmishers as the 95th . . . They could do the work much better and with infinitely less loss than any of our best light troops. They possessed an individual boldness, a mutual understanding and a quickness of eye in taking advantage of the ground, which taken altogether, I never saw equalled. They were, in fact, as much superior to the French voltigeurs as the latter were to our skirmishers in general. As our regiment was often employed in supporting them, I think I am fairly well qualified to speak of their merits.[9]

At the renewal of hostilities between Britain and America (1812–14), British riflemen took part in the battle of New Orleans.

Poor tactical handling meant the riflemen could do little to affect the outcome at this resounding British defeat. While British forces marched blindly towards well-defended enemy lines, they were mown down by coolly directed American rifle fire. The disaster was compounded by the fact that, unknown to the combatants, peace had already been declared between America and Britain. The battle was completely pointless.

The French, while adept in the handling of light infantry, never took to the rifle or the art of sharpshooting as practised by the British and Americans. They preferred to rely instead on manoeuvre and shock action. One notable exception was an incident that took place on 21 October 1805 at the battle of Trafalgar. A French marine, Robert Guillemard, was stationed on the mizzen top of the ship *Redoubtable*, his duty to shoot down on to the decks of enemy ships.

The British fleet bore down on the Franco-Spanish line, and the *Redoubtable* was engaged by the *Victory*. As the two ships lay alongside, Guillemard looked down through the clouds of smoke to see a high-ranking officer, covered in decorations, pacing the quarter-deck of the *Victory*. The French musketeer took aim and fired, from a range of just fifty feet, his target Admiral Horatio Nelson. The lead ball smashed into Nelson's left shoulder, passing through his chest before lodging in his spine. As he fell, Nelson turned to his flag captain and said, 'They have done for me at last, Hardy . . . my backbone is shot through.' The great British admiral was led down to the cockpit where he died a few hours later, the victim of the most successful single shot in the history of sniping.[10]

While the Napoleonic Wars were raging across Europe, a Scottish clergyman was working on one of the key inventions in the development of modern firearms. A keen hunter of waterfowl, the Reverend Alexander Forsyth found that the birds could see the flash of the flint hitting the pan, giving them plenty of time to escape his shot. To solve the problem he mixed a compound of fulminate of mercury and chloride of potash, which could be detonated by the blow from the rifle's hammer, and so provide the vital spark to set off the main charge, without alerting the attention of the bird. Further refinements by other sporting gunsmiths led to the percussion cap, which was placed over a nipple leading directly to the charge in the barrel.[11]

The great advantage of the percussion-lock system was that it was enclosed. Its action was not visible and it was resistant to the effects of weather (a chronic flaw in the flintlock). Also, it set off the charge far quicker. Like so many technological advances in rifle design, it came about as a result of demands for improved performance from sportsmen rather than the military. Eventually, the British Army held trials in 1834 to assess its potential, and found it twenty-six times more reliable than the flintlock. From then on the percussion system became standard in the leading military nations, and was subsequently adapted as the priming element of the modern cartridge.

Despite the percussion cap's contribution to improved reliability, the question of speeding up the rifle's reloading process remained. The answer lay in a series of separate developments in bullet design, the most famous carried out by a French officer, Captain Claude Minié, who gave his name to the system. The Minié bullet was cylindroconical in shape, with a hollowed-out base. As it was smaller than the bore of the rifle, it could simply be dropped down the barrel. When the charge was fired the force of the expanding gases drove the bullet's hollowed-out base forward so that it locked into the grooves of the rifling.[12] The Minié rifle was adopted by the British and the French, who used it during the Crimean War (1853–56). The system was also employed on both sides in the American Civil War (1861–65).

The mid-nineteenth century was a golden age of rifle design, each development swiftly followed by another, or by a different system altogether. Many of the gunmakers' new designs would shortly be tested in the great war between the Union and the Confederacy. Improvements in gunpowder composition allowed greater consistency and power, so that by the advent of the American Civil War the line infantryman was able to put out effective fire to ranges of four hundred yards and more. For the marksman these advances in propellant and rifle design made long-range shooting possible for the first time. And with the right weapon in the right hands, conspicuous individuals could be struck down at ranges in excess of half a mile. Technology had provided the framework for the emergence of the sniper.

CHAPTER TWO

THE AMERICAN CIVIL WAR

The American Civil War (1861–65) has been called the first modern war. A whole range of technical and tactical innovations came together for the first time: railways, the telegraph, aerial observation, breech-loading rifles, machine-guns and rifled cannon. Added to this was the scale of the war: the northern states of the Union drew upon the services of three million men, the southern Confederate states enlisted nearly one million. The determination of both sides to prosecute the war to the utmost led to four years of bloody conflict and a casualty list of 714,245 killed and wounded, before the Union's superior material resources brought the Confederacy to ruin and defeat.

Arguably, the single most important tactical feature of the war was the primacy of the rifle. Ordinary infantrymen were now able to deliver accurate fire at long distances with a rapidity that would normally reduce any enemy attack in the open to a broken standstill. This was little understood in the early years of the war: troops moved in massed formations and officers led ostentatiously from the front, taking their example from the commentaries on Napoleon's wars. Only when the ledger book of casualties began to fill up did tactical practice adopt a more cautious approach to the management of troops on the battlefield. But even then there were always rich pickings for sharpshooters who knew their trade.

While all aspects of military technology advanced in the years prior to the outbreak of war, progress was not uniform. Relative to small arms, artillery lagged behind. Previously, dashing horse artillery batteries had been able to gallop to within three hundred yards of the opposing line and pour a devastating volume of canister fire into the enemy ranks. When these tactics were repeated during the Civil War, the artillerymen were shot down alongside their guns. As a result they were forced to retire to safer if less effective ranges as the battlefield increased in depth. At a distance of six or seven hundred yards from the enemy the gunners could consider themselves safe from ordinary infantry rifle fire, but they still remained vulnerable to aimed fire from sharpshooters. One of the more common stories of the war involved the duel between artilleryman and sharpshooter.

While the Enfield and Springfield rifles used by the infantry on both sides were reliable weapons, capable of hitting a target six feet by five at distances of up to five hundred yards, new types of rifle were available to the committed shooter which could more than double that range.

The hunting tradition existed in both the North and the South, but was probably more pronounced in the armies of the Confederacy – certainly they supplied some of the best crack shots of the war. Firearms were in relatively short supply in the South and the Confederacy went to great lengths to secure top-quality rifles from overseas, even though it meant evading the Union's naval blockade and paying high prices. Britain was a major supplier, and the best two rifles that slipped into the harbours of Charleston and Savannah were those made by Kerr and Whitworth. These rifles were very expensive (the Whitworth cost a staggering $500,[1] against $43 for a Sharps breech-loader) and were available only in small numbers. As a consequence only chosen men were given these weapons and competition for them was invariably fierce. Because of their scarcity no attempt was made to arm complete units with Kerr and Whitworth rifles; instead, sharpshooters operated individually or in pairs while attached to regular infantry units. Given a wide degree of freedom to range over the battlefield and fight in the manner they thought best, these Confederate sharpshooters made the transition from light infantry skirmisher to genuine sniper.

The Kerr was a muzzle-loading, long-range rifle whose powerful charge rendered its bullet lethal at distances of up to a mile. The Whitworth was also a muzzle-loader and was considered to be more reliable and even more accurate at long ranges than the Kerr. Combining the precision of the target rifle and the soldier-proofing of a military weapon, the Whitworth was a formidable piece of engineering. The iron sight was graduated up to a figure of 1200 yards, and in the knowledge that they would be used for long-range shooting, most were fitted with a 14-inch Davidson telescopic sight, offset-mounted on the left of the rifle.

Utilizing the same weight of bullet as the Enfield rifle (a heavy 530 grains), the Whitworth had a reduced bore size (.45 of an inch, as against the Enfield's .577) which improved velocity and ballistic stability. This, and its superb barrel (complete with hexagonal rifling), ensured that it was exceptionally accurate. At eight hundred yards, in good conditions, the Whitworth had a mean radius of deviation of twelve inches, sufficient to ensure a reasonable chance of a successful body shot against a man standing in the open. At a range of eighteen hundred yards a Whitworth bullet still had the power to kill, and although a mean deviation of 11.62 feet made if ineffective for shooting at individuals, it was still useful as a weapon of harassment against larger targets such as gun batteries and formed-up columns of infantry and cavalry. There are several instances of hits being confirmed at ranges of fifteen hundred yards and more.[2]

In order to become a Confederate sharpshooter, the best marksmen from each infantry regiment were entered for target shooting competitions. They were required to hit man-size boards at ranges of five hundred yards, and of these only the best would be given the prized Kerr and Whitworth rifles. Once selected the men were given further instruction in the skills of the sharpshooter. Those who failed to make satisfactory progress were returned to their units and others selected in their place. As well as being first-rate marksmen, sharpshooters were expected to be brave and resourceful; one sharpshooter commander would test his recruits by escorting them into a position under artillery fire to ascertain how well they performed in real battlefield conditions. A soldier in General Patrick Cleburne's Army of Tennessee described the sharpshooters' training and their effectiveness on the battlefield:

The men were drilled in camp, on the march, and even on the field of battle in judging distances. They would be halted, for instance, and required to guess at the distance of a certain point ahead and then measure by steps on their way. When firing these men were never in haste; the distance of a line of men, of a horse, an artillery ammunition chest, was carefully decided upon; the telescope adjusted along its arc to give the proper elevation; the gun rested against a tree, across a log, or in the fork of a rest-stick carried for the purpose. The terrible effect of such weapons, in the hands of men who had been selected, one only from each infantry brigade because of his special merit as a soldier and his skill as a marksman, can be imagined. They sent these bullets fatally 1200 yards.[3]

As skilled infantrymen, sharpshooters were well versed in the tricks and stratagems of fieldcraft. When operating in wooded areas they would take appropriate measures: '[We] pin leaves all over our clothes to keep their colour from betraying us.'[4] Sharpshooters were warned not to get within four hundred yards of the enemy but to rely instead on their superior shooting skills and keep the enemy troops at a safe distance. In some instances sharpshooters would crawl into suitable positions during darkness, conceal themselves and then fire on the enemy once the sun was up.

Officers were always a priority target for sharpshooters; the disruption of command and general dismay this caused the enemy made it a most effective tactic. In one instance the Union commander General Rosecrans issued an order that his officers should wear smaller badges of rank in order to reduce the rate of casualties being inflicted upon them by Confederate riflemen.

Although most officer casualties were of junior rank – operating in the most exposed positions – there are instances of general officers being hit by sharpshooter bullets. The most famous was General John Sedgwick, who was killed at the battle of Spotsylvania on 9 May 1864. Commander of the Union IV Corps, Sedgwick was rallying his troops while under fire from Confederate positions. Seeing his men dodging from the sound of bullets whistling past, he laughingly called out: 'What! What!

Men, dodging this way for single bullets! . . . I am ashamed of you. They couldn't hit an elephant at this distance!' A Whitworth bullet immediately struck him full in the head. A close observer recorded how 'the general's face turned slowly to me, the blood spurting from his left cheek under the eye in a steady stream. He fell in my direction; I was so close to him that my effort to support him failed and I fell with him.'[5] Sedgwick's death delayed the Union advance and was a contributory factor in the Confederate victory at Spotsylvania.

During the drive on Washington city in July 1863, Confederate sharpshooters were unknowingly presented with a particularly high-value target. Captain Robert E. Park wrote: 'The sharpshooters and the Fifth Alabama, which supported them, were hotly engaged; some of this enemy, seen behind their breastworks, were dressed in civilians' clothes, and a few had on linen coats. I suppose they were "Home Guards" composed of Treasury, Post Office and other Department clerks.'[6] Park's 'Home Guards' were in fact President Abraham Lincoln and his retinue, who had left the White House to inspect the defences around Washington. A doctor standing a few feet away from Lincoln was hit, and only prompt action by a nearby Union officer in throwing the President to the ground prevented a sudden and dramatic change in the course of Civil War history.

As the war progressed and casualties began to mount as a result of sharpshooter fire, both sides attempted to deal with the problem. One method was to set another sharpshooter to eliminate the original menace (what today would be called countersniper action). A fairly standard procedure was to raise a hat above a parapet or other such defensive position in order to draw the sharpshooter's fire and reveal his position. 'I had some sand bags removed from the wall,' wrote one Confederate officer, 'leaving two holes, at each of which a marksman with a Whitworth rifle stood ready to fire. A hat raised on a ramrod drew the fire of some Federal sharpshooter, who then would look to see the effect – and quick would come the Whitworth bullet.'[7]

More sophisticated methods of sniper location were adopted, predating the tactics used by the British on the Western Front during the First World War. At the battle of Chancellorsville

(1863) a Confederate sharpshooter was causing problems for a Union position, having shot a colonel standing alongside an artillery battery. In response, the Union troops brought up a marksman of their own. His efforts were observed by a gunner from the battery:

> First he took off his cap, and shoved it over the earthwork. Of course, Johnnie Reb let go at it, thinking to kill the careless man under it. His bullet struck into the bank, and instantly our sharpshooter ran his ramrod down the hole made by the Johnnie's ball, then lay down on his back and sighted along the ramrod. He accordingly perceived from the direction that his game was in the top of a thick bushy elm tree about one hundred yards in the front. It was then the work of less than a second to aim his long telescopic rifle at that tree and crack she went. Down tumbled Mr Johnnie like a great crow out of his nest, and we had no more trouble from that source.[8]

Artillery was a key target for sharpshooters, and gunners who advanced too far forward paid dearly for their lack of caution. One of thirteen men to be awarded a Whitworth rifle in Lee's Army of Northern Virginia in 1862, John West made these observations on his fight against the gunners of the Union: 'Artillerymen could stand anything else better than they could sharpshooting, and they would turn their guns upon a sharpshooter as quick as they would upon a battery. You see, we could pick off their gunners so easily. Myself and a comrade completely silenced a battery of six guns in less than two hours on one occasion. The battery was then stormed and captured.'[9]

Sharpshooters were often employed to defend their own artillery batteries, forcing the enemy guns to operate at longer and less effective ranges. The Confederate defence of the beleaguered Battery Wagner on Morris Island during the summer of 1863 involved Whitworth-armed sharpshooters (between fourteen and two dozen men). Advancing under cover of darkness to take up hidden positions some two hundred yards in front of their own guns, the Confederate sharpshooters maintained deadly precision fire on the besieging Union forces. One exasperated Union engineer noted that the 'least exposure above the crest of the parapet will draw the fires of his telescopic Whitworths, which cannot be dodged. Several of our men were wounded by these rifles at a distance of 1300 yards of Wagner, where prisoners informed us the riflemen were stationed.'[10] In order to protect the Union gunners and engineers, slow and laborious personal defences were constructed, including circular rope mantelets for the guns. A Union attempt to overcome the Confederates with their own sharpshooters failed, and only massive artillery bombardments were able to lessen the Whitworth fire.

The use of artillery as a means of silencing sharpshooters became common practice during the war. If it turned out to be the most successful method, it was certainly an expensive one: employing a battery of six guns against a handful of men armed with rifles was a disproportionate use of resources. If nothing else such tactics underscored the economic viability of sniping – a cost-effectiveness that continues to this day.

The fighting around Knoxville, Tennessee, in November 1863 provided a good example of the deployment of artillery in a counter-sniper role. Whitworth-armed sharpshooters had taken up a position in a tower known as 'Bleak House' and were taking their toll of Union forces seven hundred and fifty yards distant. A Union artilleryman, Lieutenant Samuel Benjamin, was instructed to eliminate the enemy sharpshooters from his position on Fort Loudon some 2500 yards from 'Bleak House'. Carefully training his Parrot gun on the target, Benjamin's first shot hit the tower, killing three sharpshooters and forcing the remainder to evacuate the building.[11] Such success was rare, however, and for the most part the balance of the conflict lay with the sharpshooters rather than the gunners.

As an élite body of soldiers fighting for long periods of time in the most exposed areas of the battlefield, sharpshooters invariably sustained heavy casualties, but their *esprit de corps* ensured that their overall performance did not suffer. In the words of one Kentucky soldier: 'Several of this corps were killed during the campaign and every one of them was at some time wounded; but whenever needed there were numerous volunteers.'[12] This view was echoed by another rifleman: 'So great a reputation did the sharpshooters enjoy, in an army with many noted fighting reputations, that when one of the sharpshooters fell in action there were many who were anxious to fill the vacancies.'[13] Kerr-armed sharpshooters operating in front of the Orphan Brigade during the fighting around Resaca, Georgia, drew this comment from a Kentucky infantryman:

> Their terrible rifles soon attract the fury of the Federal artillery. Before the sun set, about half of the sharpshooters lay dead or wounded. Jim Guilliam, a sharpshooter, only left the line when a fragment of enemy shell left his right arm dangling from his shoulder only by a thin tangle of skin, flesh and bone. He walked unaided to the surgeon and underwent amputation without benefit of any anaesthetic.[14]

Overall the Confederate sharpshooters were highly regarded. Their skill at arms and readiness to enter the fray earned them the respect of the majority of their fellows. One veteran wrote: 'The Whitworth men on the Confederate side were a class quite to themselves.'[15] As for comparisons between North and South, the views of an Englishman who served with a Confederate artillery battery are not without interest – taking into account a very obvious element of partisan bias:

> Did you ever see any of those globe or telescope-sighted rifles, exclusively used by Berdan's battalions of sharpshooters in the Federal Army? They are a very accurate weapon, but expensive, I am told; yet the Federals have not done much mischief with them. The men are trained to climb trees, lie on their backs, crawl rapidly through grass, have grass-green pantaloons to prevent detection, etc.; but with all the usual systematic boasting regarding them, our Texans and others are more than a match for them. To believe their reports, nearly every general in our army has fallen under their 'unerring aim'. The best sharpshooters with us are to be found among the Missourians, Arkansans, Mississippians and Alabamians – men accustomed to woods and swamps and to Indian warfare.[16]

On the Union side, sharpshooters may never have mastered the same level of long-range shooting ability as their Confederate counterparts, but they achieved greater fame as a result of the activities of Colonel Hiram Berdan and his two Regiments of Sharpshooters. Berdan, a noted marksman and gun designer, was a complex and controversial character: his obsession with self-publicity and his ability to acquire friends in high places ensured that the repeated charges of incompetence and cowardice made against him by fellow officers did not unduly interfere with a successful military career. Despite Berdan's personal failings, he none the less played an important part in the development of sharpshooting in America.

At the outset of war, Berdan argued for the creation of a corps of sharpshooters, chosen from among the best marksmen in the northern states. Having secured the necessary authority to raise his force, Berdan set about recruitment through a series of open shooting competitions. He later wrote that the marksman would pass the initial test if he could, 'at 200 yards, put ten consecutive shots in a target, the average distance not to exceed five inches from the centre of the Bull's eye'. Berdan's intention was to train his men so that on a battlefield they could gain a first-time body hit at 220 yards and 'hit him two out of three times at a quarter mile, and three out of five times at half a mile'.[17]

The recruits poured in, encouraged by posters that promised increased pay, special service and the prospect of adventure and glory. In time the men were organized into separate battalions, the 1st and 2nd Regiments of Sharpshooters. Uniformed in dark-green coats the Sharpshooters struck a dashing figure, and even before they had seen action they regarded themselves as a cut above the rest.

While the military authorities of the Union worked out how best to use the Sharpshooters, Berdan set about finding a suitable rifle for his men. By far the most accurate weapons available were the heavy target rifles used in competition shooting. Fitted with telescopic sights (often running the length of the barrel) and weighing anything up to 30lb they were very accurate (as much if not more so than the Confederate Whitworths) but their weight, varying calibres and general lack of soldier-proofing made them unsuitable for general military use. Berdan placed orders for the five-shot Colt repeating rifle before settling on the Sharps single-shot breech-loader.

The choice of the Sharps rifle was the major determinant in deciding how Berdan's men operated. An exceptional military weapon, the Sharps lacked the long-range accuracy of a target rifle but made up for this in sheer fire power. The Sharps NM1859 breech-loading rifle utilized a sliding breech-block operated from an extended trigger guard/lever which allowed rapid fire using .52in-calibre linen and skin cartridges. The advantage of using this rifle in battlefield conditions was highlighted by a Sharp-shooter veteran: 'Being armed with breech-loaders, we could lie low, and without changing position, reload and fire ten shots a minute. A regiment of Sharpshooters in line could play havoc with an approaching column, as was afterwards demonstrated. The superiority of breech-loaders over muzzle-loaders was plainly manifest.'[18]

For the most part the Sharpshooters operated as skirmishers, in a manner similar to the British Army's 95th Rifles in the Peninsular War: crack light infantry fighting ahead of the main line. Unlike the Baker-armed 95th, however, Berdan's men had a weapon that could devastate an enemy attack, firing at a rate between five and ten times faster than the British riflemen. And in contrast to the Confederates working alone or in pairs, the Union Sharpshooters were deployed ahead of the main force in company and at times battalion strength.

Artillery remained a prime target for Berdan's Sharpshooters, demonstrated at an action at Malvern Hill on 1 July 1862. A battery of the famed Richmond horse artillery galloped forward to unlimber in front of the Union lines, where four companies of Sharpshooters were deployed. The gunners were met with a withering hail of fire which stopped them dead in their tracks, as one survivor recorded: 'We went in a battery and came out a wreck. We staid [sic] ten minutes by the watch and came out with one gun, ten men and two horses, and without firing a shot.'[19]

The Sharpshooters played a prominent role in a number of the great battles of the Civil War, notably Chancellorsville and Gettysburg (1863). At Chancellorsville both regiments fought together and virtually destroyed the 23rd Georgia Regiment, earning them official praise 'as one of the best organizations in the volunteer service'. At Gettysburg the 2nd Regiment helped defend the Little Round Top against Longstreet's repeated attacks, although in typical fashion Berdan claimed much of the credit for 'turning the tide of the war' when caught up in a minor reconnaissance action earlier in the battle.

In spite of the general tendency of Berdan's Sharpshooters to act in the skirmishing role, target rifles were still kept for 'special duties', similar to those undertaken by the Confederate Whitworth men. Indeed, some of Berdan's soldiers refused to exchange their target rifles for the Colts and Sharps, breaking away to form their own independent unit of sniper/sharpshooters. Realizing the value of high accuracy, the arms manufacturer Morgan James was commissioned to produce a target rifle for military use by Union sharpshooters. Within Berdan's corps, Sergeant Wyman S. White of the 2nd Sharpshooters explained the situation:

> Heavy target rifles with telescopic sights were used throughout the war for 'special sharpshooting'. There were not many of them, but they were assigned to those soldiers that were considered the best shots. Each gun had a special wooden case, and when the unit moved, it was carried by a supply wagon. When the man put his telescopic rifle away, he took up his Sharps rifle again, and moved with the troops until a special duty required the use of a long-range rifle again. He often operated as an independent marksman in various parts of the line where he thought he could do most good.[20]

An example of the effectiveness of a lone target-rifle marksman was provided by Private George H. Chase during an action in the Peninsular Campaign. Crawling ahead of his line Chase was able to prevent a Confederate field gun from operating. Every time the gunners tried to load and lay the piece, the Union marksman would drive them from the gun with well-aimed fire, a feat of skill and endurance that Chase kept up for two days. On another occasion, Sergeant William G. Andrews used his telescopic-sighted rifle to kill a particularly troublesome Confederate sharpshooter who had taken up a cunningly concealed position alongside a chimney in a ruined house.[21]

Things did not always go the way of Berdan's men in these sniper duels. At Yorktown, during the Peninsular Campaign of 1862, Private John Ide of the 1st Regiment engaged in a battle with a Confederate sharpshooter at long range. Armed with a telescopic target rifle, Ide was firing from the side of an old outhouse, exchanging several shots with his adversary. Soldiers from both sides looked on with interest as the duel progressed. When Ide brought up his rifle to fire again, he suddenly spun round, hit by a Confederate bullet square in his forehead. The great shout of triumph from the enemy lines only served to remind Berdan's Sharpshooters of the dangers of their calling.

Like the Confederates, Union sharpshooters were men with an independent cast of mind. Among them was Private Truman Head, better known as California Joe, an acknowledged 'character' and a first-rate scout and sharpshooter. His aptitude was demonstrated from the outset during fighting at Munson's Hill on 29 September 1861. As night fell and rain poured down on the Sharpshooters, an officer rode up and ordered them to attack troops firing from a nearby wood. California Joe strode up to the officer and exclaimed: 'You damned fool! Do you want to charge our own men?' Following a heated exchange, Joe went into the wood and brought back a Union soldier, explaining that he knew they were not enemy simply from close observation of their gun flashes, which revealed the distinctive profile of their caps.[22]

Despite the heavy casualties sustained by sharpshooters on both sides, there was always a flow of recruits to such units. Their glamorous fighting reputation ensured enthusiastic interest from the inexperienced who had yet to learn Grant's terse dictum that 'war is hell'; and for veteran troops they were a means to escape the boredom and tyranny of life in the line regiments. Over the course of the war the sharpshooters began to regard themselves as a freemasonry, separate from the ordinary soldier by virtue of the skills they had mastered and the deprivations they endured. This attitude was, at times, extended to their opposite numbers. At Gettysburg, one of Berdan's officers observed how captured Confederate sharpshooters 'expected to be hung as snipers', until they realized they were prisoners of Berdan's Sharpshooters.[23]

By and large, however, sharpshooters were justified in worrying about their fate when captured; mercy was rarely shown to an invisible hand that killed so precisely and coolly – a 'tradition' that continued into the twentieth century. On questioning Union prisoners, one Confederate soldier from the Orphan Brigade learned that, as a result of the deadliness of Confederate sharpshooters, 'their troops were exasperated, and would kill a captured man if he was found with a Kerr rifle'.[24] Whether this was in fact the case is not known, but certainly sharpshooters wore down the psychological reserves of the enemy, causing fear, anger and frustration.

And even though the skill and bravery of sharpshooters earned the respect of many, there was a general feeling that they did not play fair, taking advantage of the burdens endured by the troops on both sides of the line. A Union infantryman recalled the unwritten code of honour that forbade soldiers firing on an enemy when 'attending to the call of nature, but these sharpshooting brutes were constantly violating that rule'.[25] Another Union soldier succinctly expressed a revulsion felt by many: 'I hated sharpshooters, both Confederate and Union, and I was always glad to see them killed.'[26] As the Civil War drew to a close, so the ambivalent picture of the sniper was completed: a figure halfway between an élite infantryman and a common murderer.

In much the same way that the Gatling gun prefigured the Maxims and Vickers of the First World War, so the Civil War sharpshooters were the predecessors of the snipers fighting in the trenches on the Western Front. Typically, however, the lessons of the American Civil War went largely unheeded by the armies of Europe – they would be forced to learn the hard way. And between these two great conflicts a series of improvements was made to the rifle which would dramatically enlarge the tactical possibilities open to the sniper.

CHAPTER THREE

SNIPING AND NINETEENTH-CENTURY TECHNOLOGY

The progress in weapon development that had preceded the American Civil War gained pace in the years following the conflict. During the second half of the nineteenth century a series of inventions and improvements led to the introduction of the modern rifle. Since then – with the exception of the hybrid assault rifle – there have been only minor design modifications. A sniper of today is armed with the same type of rifle, albeit more accurate, that his predecessor would have carried in 1900.

Among the many rifle designs that had come to the fore during the Civil War, the breech-loader signalled the way ahead. For all its accuracy, the muzzle-loading Whitworth represented a design cul-de-sac; its great expense and slowness of loading compared unfavourably with the cheaper and faster-firing breech-loaders, which were steadily improving in accuracy.

The Sharps breech-loader had achieved fame during the Civil War, largely as a result of its association with Berdan's Sharpshooters. Less fortunate were the rifles made by Christopher M. Spencer and Benjamin T. Henry, which had been blocked by a conservative US Ordance Board until the latter stages of the war. They were both excellent weapons and had the great advantage of being repeaters. The Spencer carbine proved highly popular with mounted troops, but in the postwar years the design was overtaken by the Henry rifle which was marketed by Oliver Winchester, and from him took its well-known name.

The Winchester's most significant feature was its .44 rim-fire metal-cased cartridge. Not only was it more reliable and easier to fire than its waxed paper competitors, its greater durability allowed it to be housed and mechanically moved from magazine to breech within the rifle itself. In the Winchester the cartridges were loaded into a tubular magazine running underneath the barrel; a simple forward and back movement of the extra large trigger guard acted as a lever to bring the round from the magazine into the breech. The lever-action tubular repeater rifles manufactured by Winchester were an almost immediate success with frontiersmen opening up the plains of the American West, and like the Colt 'six-shooter' they subsequently entered into Hollywood cowboy mythology.

On a military level, 37,000 Winchesters were sold to the Turkish Army and used at the siege of Plevna in 1877, where the Turkish repeating rifles caused wholesale slaughter among the masses of attacking Russian troops. Despite this battlefield endorsement, lever-action rifles were not popular with the armies of Europe, who preferred the bolt-action models which were entering the market in increasing numbers. Easier to operate from a prone position, the bolt action was considered more reliable for military use.[1]

The first effective bolt-action rifle was manufactured by the Prussian gunsmith Johann von Dreyse in 1837, using a long needle-like firing pin to detonate the primer, positioned midway between the charge and the bullet. Dreyse's needle-gun was taken up by the Prussian armed forces and was successfully used by them in the wars against Denmark (1864), Austria (1866) and France (1870–71), confirming Prussia's position as the leading power in Europe. The French had followed the Prussian lead and introduced an improved bolt-action system in their Chassepot rifle of 1866.

The Dreyse and Chassepot rifles were fundamentally sound designs, but they used simple paper or linen cartridges which were hard to load and did not give a good gas-tight seal at the breech. This weakness was rectified with the introduction of metal cartridges. Developed first in France, they incorporated a primer, positioned within the base of the cartridge, which was struck by a simple hammer action, obviating the necessity for a

27

needle to pierce the charge. American gunsmiths took to the idea; one of the first designs was the .22 Smith & Wesson rimfire cartridge of 1857, followed by larger cartridges for the Spencer and Winchester rifles. An improved centre-fire priming system developed by Colonel Boxer in Britain in 1867 became standard on both sides of the Atlantic, with the cartridge cases manufactured first from coiled and then later drawn brass.

The cartridge had become the key factor in rifle design: housed within a metal case, the weight and type of both propellant and bullet would determine the performance of the weapon, bearing out the gunsmith's adage that first you design the cartridge, and then build the rifle around it.

For reasons of economy, the armies of America and Britain converted their vast stocks of muzzle-loaders to breech-loading systems. The US Springfield was fitted with a movable breechblock to become the .45–70 model, while the British Enfield was modified with a hinged breech-block designed by the American gunsmith Jacob Snider, to become the Snider–Enfield. The Snider–Enfield was never more than a stop-gap measure, however, and in 1871 it was replaced by the Martini–Henry. This rifle's action – a falling-block system – utilized a design by the Swiss gunsmith Frederich von Martini, combined with a barrel designed by the British gunsmith Alexander Henry.[2] The .45in Martini–Henry was used in the many colonial wars fought by the British in the 1870s and '80s, against Afghans and the Pathans on the North-West Frontier, and in the wars against the Zulus and the Fuzzy-Wuzzies in Africa. The Martini–Henry was popular with the troops, who considered it to have a good stopping power, a useful consideration when faced by a charge of desperate tribesmen.

The Martini–Henry can be considered as an example of a first-generation modern rifle:[3] large calibre, heavy bullet, low velocity and single shot. Other examples included the American Remington (introduced in 1867), with a calibre of 12.11mm (.513in), bullet weight of 346 grains and a muzzle velocity of 1293 feet per second, and the Russian Berdan of 1868 (a design produced by the former Sharpshooter commander), with a calibre of 10.75mm (.42in), bullet weight of 370 grains and a muzzle velocity of 1440 fps. Manufactured to high standards – as a result of improvements in machine tooling – these rifles were reliable, efficient and relatively easy to use, enabling inexperienced soldiers (by now, forming the bulk of the vast European conscript armies) to perform creditably.

While breech-loaders and metal cartridges helped conscripts master the basics of shooting, they were less important to the sniper or sharpshooter than the next key invention in rifle development. For centuries the standard propellant had been black powder, but during the mid-nineteenth century new charges were invented which were considerably more powerful and were almost entirely 'smokeless'. Experiments conducted with propellants derived from nitrocellulose and nitroglycerine were commercially exploited in the 1880s with the introduction of Alfred Nobel's ballistite and the French Army's *poudre B.*

The increased power of cartridges with smokeless charges provided higher velocity and greater accuracy at long ranges. Whereas five hundred yards had been regarded as a maximum distance for black powder shooting by the ordinary soldier, smokeless powders doubled the range. Reduced calibres and bullet weights heralded the second generation of modern rifle design. Thus while the 1871 Mauser (first generation) had a calibre of 11.15mm (.45in) and fired a bullet weighing 370 grains at a muzzle velocity of 1420fps, the second generation Mauser of 1888 fired a 7.92mm (.31in) calibre bullet weighing 154 grains at a muzzle velocity of 2864fps.[4]

The greater velocity of smokeless powders proved too rigorous for soft lead bullets: the extreme barrel temperatures led to bullet deformation and break-up during flight. As a result, bullets were either made of a harder material (such as the copper-zinc alloy used by the French Army) or, more usually, they were coated with a gilding of a stronger metal such as steel or a cupronickel alloy.

The high velocities of the modern bullet inflicted devastating explosive wounds at ranges under five hundred yards, but some soldiers claimed that the bullets simply passed through their targets at longer ranges, delivering insufficient body damage. To increase the severity of wounds at longer ranges the British experimented with dum-dum or soft-nosed bullets before adopting the Mark IV which had a hollow point that caused it to expand and break up on impact. Such bullets were, however, banned from military use by the Hague Convention in 1899 as part of a worthy if ultimately unsuccessful attempt to make warfare more 'humane'. (This proscription remains today, with the anomaly that police and other paramilitary marksmen are permitted to use hollow-nosed bullets, while the military sniper is not.) As a consequence of the Hague Convention, a compromise bullet was adopted by most armies, based on the German *Spitzer* design. The British Mark VII bullet used during the First World War was a *Spitzer* fitted with a fibre (or aluminium) tip. Although the fibre tip was primarily a device to save weight, it had the effect of making the bullet tumble in the body, thereby increasing the severity of the wound.[5]

For the sniper, the new propellants gave more consistent accuracy at long ranges; the higher velocities allowed a flatter ballistic trajectory and decreased the effect of side winds. This reduced the 'chance' factor which in the past had made long-range shooting such a (literally) hit and miss affair (it is worth remembering that for all the astonishing feats of marksmanship recorded in conflicts such as the American Civil War, there were many times more misses, invariably going unrecorded).

The new propellants also made the 'smoke of battle' a thing of the past; soldiers were no longer enveloped in clouds of white smoke once the shooting began. Previously, a single shot marked the sniper's position; now only the muzzle flash was visible, and a well-concealed sniper could normally hide the flash. This factor alone gave the sniper an enormous advantage: he was able to roam the battlefield as a near invisible assassin, producing a terrorizing effect on the morale of enemy troops.

The final improvement to the military rifle was the adoption of a magazine. Tube magazines were adapted for bolt-action use, as demonstrated in the 1884 Mauser and the French Lebel rifles, but there were a number of drawbacks, notably the weakening of the magazine spring and the possibility of bullet distortion. An obviously superior system was the box magazine developed independently by James P. Lee in America and Ferdinand von Mannlicher in Austria. The box magazine was a simple, detachable container holding anything from five to ten rounds. Fitted to the underside of the rifle, directly below the breech, the system had the advantage of allowing one magazine to be swiftly replaced by another during periods of sustained rapid fire. The box magazine was soon adopted throughout Europe and America, in such rifles as the British .303 Lee-Enfield and the American .30 Springfield.

The first major test of the new second-generation rifles came in the Second Boer War (1899–1901), between the Mauser-armed Boers and the British armed with Lee-Enfields. Although the rifles were broadly similar in design and performance, different tactical usage gave great advantages to the Boers. In the British Army the smallest tactical unit was the 25-strong section, and fire orders were given by volley to each of the unwieldy sections. By contrast, the Boers fired as individuals, which allowed them to demonstrate their considerable skills as marksmen. Also, whereas British tactical doctrine encouraged offensive action, the Boers preferred to fight from positions with cover, making good use of the defensive qualities of their rifles.

The famous Boer commando leader Deneys Reitz observed that he and his men were able to engage formed bodies of British troops at ranges of up to twelve hundred yards. At the battle of Spion Kop, Reitz wrote of a firefight between the Boers and the British, who on this occasion were holding a defensive position: 'Time after time I saw soldiers looking over their defences to fire, and time after time I heard the thud of a bullet finding its mark and could see the unfortunate man fall back out of sight, killed or wounded.'[6] The primacy of the Boers in long-range shooting brought forth a piece of advice given by old hands to raw British troops sailing for South Africa: 'Keep well clear of officers and white rocks.' Officers were readily distinguishable by their uniform and any white or pale coloured rock made an effective ranging mark for the Boer marksmen.[7]

Not every encounter went the way of the Boers, however, who sometimes found themselves outwitted by British and colonial forces. During the siege of Mafeking, the besieged British came under constant and unsettling artillery fire from the surrounding Boers, whose guns were extremely difficult to locate against the background of the high *veldt*. In an attempt to rectify this, the best shots among the Rhodesian Regiment and the South African Police were encouraged to take positive countermeasures. At nightfall, one of these marksmen would slip out of the defences, equipped with a rifle, entrenching tool and a fawny-green window blind taken from a local residence. Before daylight, he had crawled out to a position overlooking the Boer lines and dug a hole in the ground which he used as cover, the blind providing overhead concealment. The sniper waited patiently through the heat of the day until evening, when the sun was behind him and in the Boer gunners' eyes. Cautiously raising himself from his hole the sniper began to pick off his targets in the remaining minutes of good light, the Boers unable to locate the source of their discomfort due to the blinding rays of the setting sun. Before the Boers could mount a search operation, darkness had fallen, enabling the British marksman to return safely to base.[8]

Despite these local successes it was clear that British shooting standards were poor; the Boers' ability to handle their rifles only served to underline the need for good marksmanship among the British rank and file. This lesson was taken to heart, so that the British Army which went to war in 1914 possessed unrivalled skills in musketry. Unfortunately for the British – and unlike the Germans – this training had not been extended beyond simple marksmanship to include the full range of abilities called upon by a sniper.

The technological advances in weapon design made during the nineteenth century paradoxically worked against the sniper as well as for him. Whereas in the battles of previous wars armies fought as large bodies of formed men in the open, by 1900 increases in firepower made this no longer necessary or feasible; indeed, to do so invited almost certain annihilation. Troops made use of the terrain's protective features at all times, digging entrenchments whenever natural cover was absent. In the period before the outbreak of the First World War military commentators began to talk of the 'empty battlefield', where soldiers fought in small groups dispersed over wide areas. No longer could a marksman level his Whitworth rifle – as had happened during the American Civil War – against the crew of an artillery field piece, clearly visible a thousand yards away. Targets had become much smaller and more difficult to hit, especially now that brightly coloured uniforms had been replaced by khaki, field grey and olive drab. Thus, despite the improved accuracy of rifles, snipers were often forced to operate at more limited ranges to ensure a kill.

Until 1914 few soldiers took sniping seriously, and only the German Army had made provision for snipers to operate with the troops in the field. The idea of fixing a telescopic sight on to a rifle was, with the exception of Germany, virtually unknown among the European armies. The telescopic sight was a sporting device, confined to the activities of big game hunters; it was as a result of hunters' demands for good optics that telescopic sights had improved in the late nineteenth century. Magnification was not particularly high (3-power was quite average) but compared to the target sights of the Civil War, those of the 1900s had a much greater field of view, thereby improving target acquisition, and better light-collecting properties, which made targets easier to locate, especially when natural light was poor. Although the Boer War (and the Russo-Japanese war of 1904–5) had provided a foretaste of the potential of the modern rifle, it was not until after 1914 that telescopic sights were mounted on to military rifles as a matter of course – and could be used for real sniping.

PART TWO

SNIPING IN THE TWO WORLD WARS

*'Certainly, there is no hunting like the hunting of a man.
And those who have hunted armed men long enough and
like it never care for anything else.'*

Ernest Hemingway

CHAPTER FOUR

THE SNIPER EMERGES: 1914–16

TRENCH WARFARE

When war broke out in 1914 Europe was divided into two armed camps. On one side were the Central Powers of Germany and Austria–Hungary, shortly to be joined by Turkey. Opposing them was the Allied coalition of France, Russia and Great Britain, to which would be added a number of other nations, including Italy in 1915 and, most significantly, America in 1917.

The strategists on both sides had assumed that the war would be a series of vast battles of manoeuvre, leading swiftly and inexorably to a decisive victory for one side and an equally decisive defeat for the other. This, of course, was not to be. The fighting was soon bogged down in the trench warfare that gave the First World War its special character. Even on the Eastern Front, where vast open spaces prevented the development of a complete trench system, the war too became a battle of attrition – a test to see which side had the national will and material resources to continue the struggle to the bitter end.

On the Western Front – the decisive theatre of conflict – the war of movement had lasted little more than two months; by the end of October 1914 the long lines of trenches that were to stretch four hundred and fifty miles from the Swiss border to the North Sea were under construction. And it was on this static battlefield that the modern sniper emerged.

Separated by the shell-scarred strip known as No Man's Land, the troops of each side peered across at the enemy. Any soldier standing up out of his front-line trenches was vulnerable to enemy artillery, mortars, machine-guns and well-aimed rifle fire. Indeed, the entire trench system was simply a response to the destructive nature of modern firearms; only by burrowing underground could men survive in this hellish landscape. With varying degrees of success the troops began to adapt to their troglodyte existence. In the new war of trench technology the Germans were well in the lead, using mortars and grenades in large numbers. They were similarly quick to introduce snipers, first-rate shots armed with service and sporting rifles fitted with telescopic sights.

· The old Anglo-Saxon misconception that the Germans were a stolid, unimaginative people had swiftly to be reviewed by the passage of events in the first few months of the war. The Germans proved themselves to be masters of improvisation, and the speed with which they developed effective sniping in the autumn of 1914 was testimony to that talent.

The idea of using marksmen armed with superior rifles and telescopic sights was nothing new in the German Army. One writer estimated that 20,000 sights were available for military use on the outbreak of war.[1] The German High Command's realization that sniping would become a tactical element of the war led to a call for sporting rifles of all types to be rushed to the front. Although most of these rifles used the same 7.92mm calibre as the German service rifle, they proved too vulnerable for front-line conditions and were replaced by military issue weapons when they became available.

The basic service rifle used by the German Army was the 7.92mm Gewehr 1898 (Gew 98), a development of the Mauser Modell 1888, the rifle used by the Boers so successfully against the British in the Second Boer War. Fitted with a five-round magazine, the Gew 98 was a superb rifle, accurate and reliable. Although the Mauser bolt-action was slower than that of the British SMLE rifle it provided greater accuracy, which, of course, was more important to the sniper.

The German authorities issued their snipers with improved (or 'accurized') Gew 98s, upon which were mounted telescopic sights, normally either 3- or 4-power in strength. Germany was at the forefront in the manufacture of optical instruments, and its snipers were well equipped with binoculars and telescopic sights. The *Scharfschützen-Gewehr 98* (Sharpshooter Rifle 98) was modified to include a turned-down bolt handle to keep the action clear of the telescopic sight. The Gew 98 would be used by German snipers throughout the First World War.

As well as sending telescopic-sighted accurized rifles to the front line, the Germans developed an improved cartridge for the exclusive use of snipers. Furthermore, guidelines were drawn up for the effective deployment of crack shots in a genuine sniper role. An extract from an official document of 1915 explains the role of a trench sniper. More than just acting as sharpshooters – good shots who were prepared to fire away at targets of opportunity – these early snipers were organized at company level to operate with complete freedom in a single task, that of killing individual enemy soldiers with accurate long-range rifle fire:

The weapons with telescopic sights are very accurate up to 300 metres. They must only be issued to qualified marksmen who can assure results when firing from trench to trench, and especially at dusk or during clear nights when ordinary weapons are not satisfactory.

The marksman must shoot with discretion and the rifles must not be fired for salute or suppressive fire. Marksmen are not limited to the location of their unit, and are free to move anywhere they can see a valuable target. Sentries have the duty to signal the marksman, such targets they themselves can determine.

The marksman will use his telescope to watch the enemy front, recording his observations in a note book, as well as his cartridge consumption and probable results of his shots.

Marksmen are exempted from additional duty. They will wear a special badge of two crossed oak leaves above the upper badge of the cap.[2]

The Germans' skill in sniping can perhaps be best appreciated in comments from those who had to suffer their attentions. Captain J. C. Dunn, the medical officer of the 2nd Battalion the Royal Welch Fusiliers, became aware of the danger of German snipers as early as October 1914. The Germans adopted well-concealed positions to fire on enemy working parties going about their trench-repair duties, the British being particularly vulnerable to attack through the absence of adequate communication trenches. The German snipers protected their own men working in exposed positions, and, 'worse still,' Dunn wrote, 'they covered attacks, preventing our men lining the parapet until the attackers were close up.'[3]

This systematic approach to sniping was totally lacking in the Allied camp. The French Army had snipers but they were not used to great effect; shooting with telescopic sights at long ranges seems to have had a low priority throughout the war. The British arrived at the front with no sniping experience. The high level of rifle-fire casualties suffered by British forces during the first two years of the war reflected the dominance of German snipers and the inability of the British to counter this threat.

The British infantryman was armed with the .303in (7.7mm) Short Magazine Lee-Enfield (SMLE) rifle, officially designated Rifle No. 1 Mk III (followed by the easier to manufacture No. 1 Mk III*). The SMLE was a good service rifle. Equipped with a ten-round magazine, its easy bolt action made possible a high rate of fire while its robust construction and relatively short length made it an ideal weapon for use in the trenches. Although some snipers found the SMLE acceptable, most preferred the more accurate Rifle No. 3 Mk I – better known as the P.14 – which utilized the slower and more secure Mauser bolt action (where the locking lugs were at the forward end of the bolt). Long and awkward to use and harder to maintain than the SMLE, the P.14 was a product of the Bisley school of rifle design, but it had a heavier barrel and a stronger receiver and was a more accurate weapon than the SMLE, making it well suited for the sniper.[4]

Designed on similar lines to the P.14 was the Canadian-manufactured Ross rifle, which armed Canadian troops when they first arrived in Belgium and France. The Ross bolt action required the most careful maintenance if it was to operate effectively and safely; there were widely reported instances of the bolt flying back into the firer's face. Once the harsh conditions of trench warfare had been encountered by the Canadian troops, they exchanged their Ross rifles for the more dependable SMLEs. But the Ross was a good marksman's rifle, and numbers remained in the Canadian inventory throughout the war for sniping purposes.

Captain (later Major) H. Hesketh-Prichard was a former big-game hunter who came out to France in May 1915, and was dismayed at witnessing the 'severe gruelling' being inflicted on the British by German snipers. As a result of these initial experiences Hesketh-Prichard became a pioneer of sniping in the British Army. He wrote of his early encounters in the book *Sniping in France*: 'It is difficult now to give exact figures of our losses. Suffice to say that in early 1915 we lost eighteen men in a single battalion in a single day to enemy snipers.'[5] Losses of this scale swiftly drained the lifeblood out of a battalion. The damage done was more than just material, as morale suffered in proportion to the level of enemy domination.

The sudden crack of a sniper's bullet, followed by a soldier falling from the parapet in his death throes was an obviously unnerving experience for onlookers. The horrifying effect of a gunshot wound to the head is poignantly described in this account by a young British officer, Charles Carrington, during an engagement on the Somme in 1916. One of his men, Corporal Matthews, had been killed by a sniper on a poorly dug traverse. An attempt was made to repair the position:

Almost as we approached and cut into it with pickaxes, the same sniper fired again from the village to our left, and a man called Pratt dropped like a stone just where the corporal had fallen. He, too, had a small round hole in his temple and the back of his skull blown away.

Pratt was beyond hope. His head was shattered: splatterings of brain lay in the pool of blood under him; but, though he had never been conscious since the shot was fired, he refused to die. An old corporal looked after him, held his body and arms, which writhed and fought feebly as he lay. It was over two hours before he died, hours of July sunshine in a crowded space where perhaps a dozen men sat in a ditch ten yards long and

five feet wide, reeking with the smell of blood, while all the
time, above the soothing voice of the corporal, a gurgling and a
moaning came from his lips, now high and liquid, now low and
dry . . .[6]

The 'severe gruelling' noted by Hesketh-Prichard was confirmed
by another sniping expert, Major F. M. Crum of the King's Royal
Rifles. Crum had service experience already but realized that
trench warfare demanded a new tactical approach, especially as
a result of German sniping prowess:

My first visit to the trenches left a lasting impression on me. It
was a good regiment to which we were attached for instruction,
and the state of affairs which prevailed was probably represent-
ative of other parts of the line. Wherever we had been in the
front trenches, the tops of sandbags were being constantly
ripped by bullets, and periscopes were being broken. Occasionally one of our men would jump up and take a snapshot over
the parapet, but for the most part our men were keeping well
down, never firing a shot; laughing perhaps a little uncomfort-
ably, as they watched the apparently lifeless trenches in front
of them through the broken remains of vigilant glasses.

At one point we crawled to an isolated trench, sniped at as
we went, wherever the communication trenches were exposed
to view. The colonel put his periscope up. It was shot at once
and he got a knock in the face. Covered in mud, he turned to
his men and said, 'We mustn't let them have it all their way'.
But neither he nor I had any clear idea how the thing was to be
stopped.[7]

The shooting up of periscopes was quite common along the front
line of the British trenches. The poet Robert Graves, then an
officer in the Royal Welch Fusiliers, relates how a German sniper
at Cuinchy had fired at his trench periscope, which was only an
inch square, and yet had '. . . drilled it through, exactly central,
at four hundred yards range'.[8]

British frustration at dealing with German sniper dominance
was ironically compounded by the BEF's fine tradition of musketry, which had served it so well in the battles of Mons (August

1914) and First Ypres (October–November 1914). The British soldier was trained to create a beaten zone in which any living thing would be killed, consequent on well-directed and rapid fire. Individual accuracy was not crucial to this type of fire by sections and platoons.

Typically, the British response to the German sniper threat was amateur and ad hoc. On their own initiative, men like Hesketh-Prichard and Crum set about 'beating the Boche at his own game'. Dependent on the support of a sympathetic battalion CO, the sniper devotees acquired sporting rifles and began to shoot it out with the Germans on the other side of No Man's Land. Throughout 1915 the efforts of these men remained unofficial, some of the funding for weapons supplied by interested civilians out of their own pockets.

An adequate supply of telescopic sights was a first requirement. They not only magnified the target but gave greater illumination during periods of poor light, whether at dawn or dusk, or when the sky was overcast. Magnification was not particularly high – typically around two- to three-power – but the lenses had good light-catching properties which provided the sniper with a clear image. A further advantage of telescopic sights lay in the fact that while a soldier would not reveal himself to the naked eye of the enemy opposite, he would be unaware of the possibilities of telescopic sights to pin-point him at long range. Thus he would adopt exposed positions confident of not being shot at, forgetting that a sniper might have him in his sights.

Some sights were fitted with dials to adjust range and windage – movement to left or right to take into account the effect of wind. But other sights were fixed and even when they were not, the sniper would often not have time to dial in the appropriate corrections. As a result a sniper had to learn to 'aim off': that is, mentally to allow for these variables and aim away from the target by the correct amount. Swift 'aiming off' was an art which only the most experienced and skilful snipers could achieve, especially at long ranges where the effect of these variables increased enormously.

Unfortunately for the British, almost all optical instruments remained in short supply, due mainly to the near German monopoly in this area of precision engineering. So great was the problem that in the autumn of 1915 the British government even put out feelers to Germany, via Switzerland, to acquire telescopic sights and binoculars. The Germans, anxious to secure stocks of natural rubber, were prepared to engage in unofficial barter and provided a detailed list of items for exchange. This dubious means of procurement was not in fact taken up, as a reasonable supply of sights began to arrive through more conventional channels.[9]

Once the sights had been acquired they had to be mounted on to the rifle, which often proved problematic because the service rifle had not been constructed for the use of telescopic sights. As a stop gap measure sights from sporting gun makers were grafted on to rifles, and although often first-rate optical instruments, they were expensive and lacked uniformity. By 1916 government orders for sights were being met, the majority being either Periscopic Prism or Aldis models.

Many of these sights were not centrally mounted, but attached to the rifle offset to the left. This had the advantage of allowing swift reloading by the standard five-round clips, which proved popular with those snipers who wanted the potential for rapid fire. But there were disadvantages: firstly, it was difficult for the firer to obtain a tight cheek-to-butt 'spot weld', so that pads had to be attached to the butt to ensure a good aim; secondly, a larger opening had to be made in loophole plates, which made the sniper more vulnerable to return fire. Although in the end the centrally mounted sight won the argument, offset mounts were still being used until after the Second World War.[10]

The mere possession of rifles with telescopic sights did not, however, make soldiers into snipers, and in many instances the British troops assigned to 'Hun hunting' had little idea how to set about the task. Indeed, at times the issue of telescopic sights to untrained men could be a positive hindrance to good shooting. The novice sniper, using sights not zeroed correctly, would become over confident through the good vision provided by the sights, and in the end hit nothing. Because telescopic sights were often considered little more than trench stores, they were handed out by the quartermaster sergeant to incoming troops along with pickaxes, shovels and rubber waders. Thus many sniper hopefuls were using sights which were hopelessly out of alignment.

British inexperience extended beyond the workings of their equipment. H. W. McBride was an officer in the US Army and a regular competitor at Camp Perry. At the outbreak of war in 1914 he joined the Canadian Army in order to see action. Arriving in France in 1915 he was struck by the amateurishness of British would-be snipers:

> One day in September I was scouting around the back of our lines opposite Messines Ridge and I ran into a sniping post, manned by an officer and two snipers from the Buffs. These fellows were using the ordinary Lee-Enfield rifles, upon which they had mounted telescopes. The nearest enemy targets were at least eleven or twelve hundred yards away, but they were doing most of their shooting at targets well over on Messines Ridge and I knew the range to these points was about two thousand yards. My personal experience has been that the firing of single shots, at individual targets, at ranges of more than a thousand yards is just a waste of time and ammunition.
>
> But this sniping outfit from the Buffs was deadly serious and I just had to admire their spirit even though their judgment was bad. I ventured to ask the officer if he thought it was worth while to shoot away at targets located a mile or more off, and he replied they were acting under orders. I came away then, and I imagine that outfit hung around there for many more days engaged in such useless work.[11]

McBride's reservations were supported by comments from British infantrymen in the trenches. Sergeant-Major Ernest Shephard was an experienced infantryman in the 1st Battalion, the Dorset Regiment. His diary entry for 16 March 1915 reads: 'One of our snipers wounded slightly about 10 a.m. In the evening we made a snipers' nest in a trench, which drew heavy fire next day, so we pulled it down.'[12] Despite such setbacks, the more enterprising elements of the British Army began to get to grips with the sniper problem. Among these was B. A. 'Nobby' Clarke, a regular soldier in the 2nd Battalion, the Northants Regiment. In the trenches around Neuve Chappelle, Clarke was appointed to the role of sniper in April 1915:

> We came out of the trenches and went to a village where we stayed eight days. During that time the RSM sent for me and Lance-Corporal Budworth. He told us four sniping rifles would arrive the next day. Orders were sent to each company to supply two marksmen or first-class shots. We received our telescopic rifles. I was made lance-corporal, and Budworth and I were to run the snipers. No officers could be spared for us; they were few enough. We made targets and set off across the fields and found a bank just right for practice. We fired off a lot of shots at four hundred yards; it was impossible to miss your target.
>
> The newly formed snipers got to work putting the sniping posts in disused trenches just at the back of our front line. We picked the highest points so that we could see right beyond his [the enemy's] front line, because observing was of much more value than just killing a Jerry or two. You tried to spot his machine-gun and artillery positions. When he fires, your observer makes notes and we mark on the map the exact position of his guns – the rest is left for the artillery to deal with. We also looked for troop movements: everything, however small, had to go in the book [sniper's notebook], and the details sent to Brigade HQ.
>
> It took the Jerries days to find we had the same kind of sniping rifles as they had. I think we sent a nice few shots home, but we only reported how many we fired during the day. As I told our RSM, when he asked how many we had killed, 'Sir, we are waiting for the Germans to send their report in.' Never be bluffed if a sniper swanks how many he has killed. None of our lads did. We knew it was almost impossible to miss but unless the enemy is shot in the open and lies quiet for a while you would not be certain.[13]

Although self-taught, Nobby Clarke and his comrades had demonstrated some of the essential requirements of the successful sniper: confidence, patience and intelligence.

*

Along with the introduction of the telescopic rifle, other aids were brought into the trenches to improve performance and safeguard the sniper from enemy fire. From the outset of hostilities the Germans had used steel plates – two to three feet in length, eighteen inches to two feet in height, fitted with a loophole in the centre – to protect their men on front-line observation duties. These plates were proof against standard-issue rifle bullets, shell splinters and shrapnel, and they soon became a standard part of the sniper's equipment on both sides of No Man's Land. Most would be left in fixed positions, although some snipers carried plates around with them from place to place.

There were two main ways of dealing with a sniper plate: to call on artillery to blast the position to smithereens – a form of extreme discouragement – or to use snipers issued with armour-piercing (AP) ammunition to kill the enemy sniper. AP bullets were not available to the British until 1917 but the acquisition of special high-velocity express rifles and large-calibre 'elephant' guns (both used in big-game hunting) proved equally effective. As a result, loophole plates – on both sides – had to be cunningly hidden behind sandbags or other pieces of cover (in addition, it was important to conceal plates to prevent the enemy from becoming aware of the presence of snipers and acting more cautiously).

British parapets tended to be carefully built with a neat and regular appearance, making movement easy to spot from the enemy side. By contrast, the Germans had deliberately avoided this danger. Their trenches were uneven in construction with sheets of corrugated iron and coils of barbed wire breaking up the outline of the parapet; sandbags of differing colours were used to confuse enemy observation further. In the words of Hesketh-Prichard, the 'Germans had a splendid parapet behind which a man could move and over which he could look with comparative impunity, whereas we in this respect gave heavy hostages to fortune.'[14] The British sniper pioneers had to fight hard to get the authorities to accept the 'untidy' trench line. Comparative trials were held between the two trench types. When a dummy head was held aloft for a few seconds the number of hits was scored in a ratio of three for a 'tidy' trench to one for an 'untidy' trench.[15]

But no matter how well the forward trenches were camouflaged, a man sticking his head over the top in a fiercely contested area virtually guaranteed himself a sudden and violent end. An early solution to the problem was the trench periscope. Even though they could be knocked out by enemy sniper fire, they allowed the soldier reasonable and safe observation, and they could be easily replaced when destroyed or damaged.

The periscopic principle was then extended to rifle fire. The rifle was placed in a simple wooden jig; a periscope was fitted to the butt plate to look along the line of the sights and a rod attached to the trigger to allow the rifle to be fired by remote control. Limited to effective ranges of a hundred yards or less the periscope rifle – or 'sniperscope' – was a basic weapon, which could be fired by any front-line occupant and drew little upon the true sniper's skills.

Like the 'sniperscope' the rifle rest and rifle battery came under the general remit of the sniper officer, but similarly they were not precision weapons and could be operated by any competent soldier. A rifle rest comprised a rifle clamped to a fixed position, its sights trained to bear on a particular position in the enemy rear, such as a new earthwork or an exposed communication trench. The rifle would usually remain hidden and unused during the day and fired only at irregular intervals during the night to catch unwary troops believing they were safe in the darkness. Consequently, hits were judged to be merely casual stray shots (a near-continuous feature of trench warfare), unlike a burst from a machine-gun, whose position would be noted and steps taken to eliminate the position. The rifle battery was a more sophisticated contraption, a wood or steel frame capable of holding up to six rifles at once, which normally allowed for some movement in elevation and traverse. Later in the war, however, the rifle batteries were replaced by machine-guns, whose gunners were trained to loose off a couple of rounds to simulate rifle fire.[16]

By mid-1915 these devices had become commonplace artefacts of war, along with rifle grenades, trench mortars, flame throwers and gas dispensers. The old divisions of 1914, composed primarily of rifle-armed infantry, sabre-wielding cavalry and horse-drawn artillery, were being transformed into a modern army of specialists. The sniper was just a part of the industrialization of war.

GALLIPOLI ADVENTURE

On the Western Front, heavy casualties and strategic impasse had led Allied politicians to look for other areas in which to prosecute the war. One such region was the Turkish-held Dardanelles, the waterway linking the Mediterranean to the Black Sea. Allied command of this vital channel would in all probability knock Turkey out of the war, as well as providing an excellent supply route to the hard-pressed Russians.

Tentative steps by British and French ships to secure the straits had foundered on the mines and artillery of the Turkish defenders. Amphibious landings by the Allies (British, Anzac and French forces) on 25 April 1915 secured only tenuous footholds, and for the rest of the year the Allies fought a bloody war of attrition against the German-backed Turks. The landscape over which the two sides fought was a pitiless tumble of ridges and gullies, devoid of water and vegetation. As on the Western Front, defenders and attackers dug in to avoid shot and shell. Here too, the sniper became a man to fear and respect.

Although they lacked the all-round training of their German co-belligerents, the Turks proved to be adept in the art of sharpshooting and sniping. Armed with Mauser rifles the Turks had the advantage of firing down at the Allies from the heights of the Gallipoli Peninsula. From the moment they had landed Allied troops were sniped at constantly: 'Put your head up out of the trench,' wrote one British officer, 'and whizz goes a bullet from an invisible sniper.'[17] The British writer and humorist A. P. Herbert was a soldier in the Royal Naval Division (RND); his 'beachcombing' experiences on Gallipoli included the horrors inflicted by enemy snipers:

In that first week we lost twelve men each day; they fell without a sound in the early morning as they stood up from their cooking at the brazier, fell shot through the head, and lay snoring horribly in the dust; they were sniped at as they came up the communication trench with water, or carelessly raised their heads to look back at the ships in the bay, and in the night there were sudden screams where a sentry had moved his head too often against the moon.[18]

The troops on Gallipoli were forced to learn the lesson of the British on the Western Front: carelessness was fatal. A trooper in the 5th Light Horse AIF, Ion Idriess, kept a diary which listed the destructive power of enemy snipers: on 26 May seventeen Australians were shot by snipers in one place; on 28 May a further fifteen were hit in the morning. Shortly afterwards Idriess recalled how sniper shots could be deadly in areas thought safe from enemy rifle fire:

He was a little infantry lad, quite a boy, with snowy hair that looked comical above his clean singlet. I was going for water. He stepped out of a dugout and walked down the path ahead whistling. I was puffing the old pipe, while carrying a dozen water bottles. Just as we were crossing Shrapnel Gully [Valley] he suddenly flung up his water bottles, wheeled around, and stared for one startled second, even as he crumpled to my feet. In seconds his hair was scarlet, his clean white singlet all crimson.[19]

The high rate of casualties caused by Turkish sniping was gradually controlled by improved defences and a greater sense of caution shown by the troops. In addition, the British and Anzacs instituted a determined counter-sniper programme, using their own snipers to deter and kill the Turkish marksmen. The close proximity of the front line allowed anyone who fancied himself a marksman to try his hand at sniping. As time went by the crack shots with impressive tallies began to emerge. And because of shortages of artillery and machine-guns, well-aimed rifle fire was more important on Gallipoli than in other theatres of war.

For some soldiers, sniping was a diversion from the tedium

of trench warfare – all part of the 'great game' of man hunting. Private A. McCowan, a rifleman in the 5th Battalion of the 1st Australian Division, regarded sniping as a pastime, although of a deadly sort:

10 May 1915 I am going out to try my hand as a sniper as I am reckoned a very tidy shot.

11 May I have been sniping with a Turkish rifle and had some very good sport. It is surprising the time you can have with a pair of field glasses and a rifle.

12 May I am out again sniping. It is an exciting game, far more than football.

11 June I was out for twenty-four hours but only added three to the list. The total now stands at seventeen that I am sure of.

15 June I was out this morning at 2.30 and obtained a few good shots. Two Turks were coming from a well with water. I got one and then wounded the other. I would be the first to say that the wounded should not be shot, but if you are too far [away] to make him surrender, are you going to let him toddle off and lose him? It's all right in theory but no good in practice.[20]

The ranks of amateur marksmen included their fair share of characters. A Sergeant Brennan, formerly of the Dublin Fusiliers, was a highly regarded cook in the field kitchens of the 7th Light Horse on Gallipoli. After cooking the men's breakfasts, Brennan would pick up his rifle and carefully move through the trenches to the front line. He specialized in tracking down Turkish sniper posts, although any Turk showing his head above the parapet was fair game. His mood would rise and fall depending on the success of his sniping operations. 'Some days,' wrote fellow soldier Oliver Hogue, 'he would come back to camp angry and disappointed. "Thirty shots and not a single scalp," he exclaimed, kicking aside some innocent mess tin. But at other times he stalked back as if he had won the battle of Anzac on his own. "Killed three Turkeys," he cried. And then he was as happy as Larry all day.'[21]

In distinguishing the mere enthusiast from the true sniper, Hogue made this telling comment:

Most of us try our hand at sniping with more or less success –
climbing up and down these hills in search of what we call 'big
game' – but although I am a tolerable shot myself, I have come
to the conclusion that your true sniper, like your true poet, is
born not made . . . It is not enough to see your man and have a
pot at him, for the chances are that just as you let go, he stoops
down to pick a pretty flower, or he stumbles over a scrub root.
Now, the successful sniper is he who anticipates that stumble,
or with an uncanny sort of second sight sees that pretty flower
which the enemy gentleman is going to pluck, and aims low
accordingly. Only by some sort of intelligent anticipation could
some of our men put up the astonishing records that stand to
their credit.[22]

The most distinguished Gallipoli sniper was Billy Sing, a Queens-
land native and a trooper in the 5th Light Horse. Before the war
Sing had been a crack shot, a regular prize-winner in shooting
competitions at Brisbane and Randwick. But he also understood
the disciplines of sniping, particularly patience. Rather than just
blast away at an upraised head, he made sure that when the
trigger was pulled his bullet would find 'its billet'. Sing devoted
himself exclusively to sniping, and preferred to work with a
comrade who acted as an observer, common practice later in the
war but something of a rarity then. 'A little chap, very dark, with
a jet-black moustache and goatee beard,'[23] Sing had an extraordinary 150 confirmed kills to his name on Gallipoli. His single-
minded determination is revealed in this passage from Oliver
Hogue:

He played the Turks at their own game, and beat them badly.
He himself admitted it was 'a shame to take the money'. He
used to sit with his rifle set at a certain track which the enemy
thought was well concealed behind the hills. His mate had a
telescope and spotted for him. They waited till they saw a head
appear [but] Sing only grinned and waited. He would get his
Turk later on. Emboldened by fancied immunity, the unsuspecting one would show his head again, then his shoulders,
then half his body. That was quite enough. The spotter said
'Right'; the rifle fired and another victim fell.[24]

The talent for sniping shown by Anzac forces was a reflection of their rigorous outdoor background, where hunting and especially bush shooting were commonplace. The dictum that hunters made the best snipers held good for Australia and New Zealand, as it did for Canada, South Africa and the United States.

In the Canadian Corps there were a number of big game hunters who quickly took to sniping. Their experience in stalking and using telescopic sights ensured that they were virtually trained snipers when they arrived in France. These men gained immediate parity with their German opposite numbers and soon achieved sniping mastery. Major (later Lieutenant-Colonel) N. A. D. Armstrong had been a crack shot in British Columbia prior to his arrival as an instructor at the 2nd Army's Scouting, Observation and Sniping (SOS) School (he was subsequently appointed Commandant of the Canadian School of SOS in 1917–18). Armstrong intelligently combined theory and practice to produce a first-rate crop of Canadian snipers.[25] Among these was Private P. Riel, a French Canadian in the 1st Canadian Division, who shot thirty Germans between March 1915 and mid-January 1916, before being killed by shell-fire.[26]

South African snipers achieved some spectacular scores on the Western Front, including a tally of over a hundred kills for Lieutenant Neville Methven. The South Africans who were sent to France were part of a private initiative financed by the wealthy businessman Sir Abe Bailey. The eighteen hand-picked crack shots went under the name of 'Sir Abe Bailey's South African Sharpshooters' and arrived at the front in 1916 where they served with the British 1st Division. Under Methven's command the South Africans used accurized SMLE rifles with telescopic sights to considerable success, although they paid a high price: of the original eighteen that left South Africa, only six returned.[27]

In contrast to the Commonwealth, rifle shooting in Britain was confined to the target ranges, where the vital sniping requirement of 'stalking the prey' was missing. The only exception was in the Highlands of Scotland; there the ghillies' (gamekeepers) experience of stalking deer over long distances proved invaluable for sniper training. This was not to say that the British made poor snipers, only that there were fewer natural shots to choose from and that more training was necessary to bring them up to a high standard.

VIOLENCE AND MODERATION

Although the First World War appeared to fulfil the Clausewitzian concept of total war, various influences came into play to reduce the otherwise terrible levels of violence. During an offensive, armies threw all their efforts into destroying the enemy; at other times the war was conducted in a more circumscribed manner. Without this moderating factor the material and – especially – the human resources of an army would have been swiftly exhausted.

At the front, truces were fairly common. Apart from the famous Christmas truce of 1914, there were many localized cease-fires, so that the wounded could be brought in from No Man's Land, or running repairs made to trenches during bad weather. In quiet sectors of the line there would be unwritten agreements to limit conflict to more 'reasonable' levels, so that, for example, soldiers would fire their weapons to 'keep up appearances' but would deliberately aim high. Faced with the shared misery of trench existence, the front-line soldiers of both sides often preferred to adopt a 'live-and-let-live' philosophy.[28] Such cosy relationships were, however, easily and regularly disrupted by the onset of an offensive or the arrival of an aggressive unit. Snipers were generally a force for undermining the live-and-let-live system, although in practice their actions were often influenced in favour of moderation.

The sniper's prime function – to kill enemy soldiers – invited retaliation, and this was not welcomed by the sniper's infantry comrades. While the good sniper would take up a new, safe position far away from the scene of his work, the enemy would call on mortars or artillery to destroy the trenches where his old position had been – causing casualties and distress to the regular occupants. Accounts by snipers invariably refer to the antipathy directed at their activities by their own troops.[29]

As early as December 1914, Captain J. C. Dunn of the Royal Welch Fusiliers had noticed the workings of the live-and-let-live system. Referring to a high incidence of German sniping, Dunn wrote: 'Not all of it was in deadly earnest. On the left the Germans amused themselves by aiming at spots on the walls of cottages and firing until they had cut a hole. Our men said, "Fritz had to be chummy, he had a brewery working just behind him".'[30] The obvious inference being that over-accurate German sniping would have brought retaliatory British artillery fire on a valuable asset.

The moral repugnance felt in striking down a soldier in cold blood stayed some men's hands, but such feelings were alien to the true sniper. While sniping at long range, Robert Graves looked through his telescopic sights at a German taking a bath in the third line. Graves had qualms about shooting such a defenceless target; he neatly sidestepped the issue by giving the rifle to his sergeant, who duly despatched the careless bather.[31] Victor Ricketts, a sniper in a territorial battalion of the Gloucestershire Regiment, provided the standard line on the morality of sniping: 'It's not too pleasant to have a fellow human in one's sights with such clarity as to be almost able to see the colour of his eyes and to have the knowledge that in a matter of seconds, another life will meet an untimely death. However, one had to be callous; after all it was an eye for an eye, a tooth for a tooth.'[32]

For the determined sniper, the prospect of enemy retaliation and the curses of one's comrades were overridden by the thrill of the hunt. H. W. McBride, the American soldier serving with the Canadians, offered a robust defence for the aggressive spirit when questioned by the ordinary line infantry: '"No!" they would shout. "Don't do that. They will retaliate." Well, hell's bells, let 'em. What the Devil are we here for? A summer picnic? To me it was a great game. Whenever they came back with their retaliation I was just as pleased as a schoolboy who has received the highest possible grade. It was proof positive that I had stung them.'[33]

A less bellicose account from the writer Ian Hay describes the role of the sniper in maintaining the war during lulls in the fighting:

On an off [quiet] day . . . the sniper is a very necessary person. He serves to remind us that we are at war. Concealed in his own particular eyrie, with his eyes forever laid along his telescope sight, he keeps ceaseless vigil over the ragged outline of the enemy trenches. Whenever a head, or anything resembling a head shows itself, he fires. Were it not for his enthusiasm, both sides would be sitting in their shirtsleeves upon their respective parapets, regarding one another with frank curiosity; and that would never do.[34]

Alongside trench mortars, rifle grenades and night patrols, snipers were used to ensure the domination of No Man's Land. In a war without the short but decisive battles of previous conflicts, success in the front line was often measured by a battalion's control of No Man's Land, which for first-rate units became an end in itself. When entering the trenches at Cambrin in May 1916, the Second Battalion of the Royal Welch Fusiliers reckoned it a point of honour to wrest control from 'the pugnacious regiment opposite'. Over a four-day period the Royal Welch battled with the Germans, and in Dunn's words: '. . . the close array of German periscopes was thinned, and we had observation of his parapet . . . Counting periscopes as points, the Battalion was easily the winner.'[35]

When run-of-the-mill units were replaced by an aggressive battalion, the snipers had an immediate advantage in finding German targets grown careless through previous inaction. Victor Ricketts became aware of this when his battalion of Glosters took over a stretch of trenches from the French in the summer of 1915: 'The French, it seems, were prepared to let things remain quiet, preferring to make a quick sortie, then lie low. This resulted in our being presented with many targets by unsuspecting Germans . . .'[36]

On both sides of the line the sniper had acquired an élite status, and was a respected if unpopular figure with his comrades. For men who had the requisite skills, the life of a sniper held certain advantages. Spared the drudgery of ordinary soldiering the sniper held a roving commission; operating independently within his battalion he was able to exercise his own initiative – an opportunity unknown to most private soldiers.

On the other hand the sniper was more exposed to danger than the ordinary soldier, although this could be lessened by guile and good fieldcraft. The greatest danger lay in capture, the general consensus being that there would be no mercy shown to the sniper. On the British side it would seem that troublesome snipers were usually killed out of hand.[37] Charles Carrington's unit had suffered from sniper fire during the Third Battle of Ypres (1917): '. . . we barely noticed . . . the four snipers who had held us up so long [had] slipped into the crowd of captives and went away with them. We should certainly not have given them quarter if we had thought of it in time.'[38]

Through 1915 and into 1916 British and Commonwealth soldiers slowly overcame German sniper superiority. As in other areas of military endeavour the British Army was becoming steadily more professional. Improvements in British sniping were largely a result of the work of men such as Hesketh-Prichard, Crum and Armstrong, who had instigated a policy of thorough training at battalion level throughout the Army. Gradually a body of proficient snipers entered the trenches where they could do most harm. These basic advances were further developed and refined during the remainder of the war years, as sniper training and deployment was given official status.

CHAPTER FIVE

A FULLY-FLEDGED ARTICLE: THE SNIPER 1916–18

THE NEW PROFESSIONALS

By the middle of 1916 the First World War had become a gigantic slogging match; each side tried to grind the other down, as much by the brute application of economic muscle as by conventional military means. The old arts of improvisation were superseded by a belief in the virtues of the system, what the soldier-poet Edmund Blunden called the transformation 'from a personal crusade into a vast machine of violence'. In the area of sniping, as in other specialist military fields, this signalled the demise of the inspired amateur and his replacement by men trained to a uniform standard. And if the flexibility of the old informal system was lost, then there was compensation in the form of a new and much needed professionalism – which in the end yielded better results.

During 1915 and 1916 sniper training in the British Army was slowly and somewhat erratically extended throughout Flanders and France. SOS (Sniping, Observation and Scouting) Schools were set up on an army-wide basis, the First Army, according to Major H. Hesketh-Prichard, being in the vanguard of sniper training. As chief instructor to the First Army SOS School, Hesketh-Prichard's comments must be taken as somewhat partisan, although he claimed that it 'turned out three snipers to any other Army's one'. By the end of 1915 he estimated that at least one officer from every battalion in his area of responsibility had attended a sniper course. But it was not until 24 November 1916 that the school became official with its own establishment. The essential function of the SOS schools was to produce sniper instructors who, after graduation, would go back to their own units and organize and train sniper sections in the front line.

Initially, training concentrated on improving marksmanship but over time the curriculum extended to fieldcraft, camouflage, map-reading, use of the compass, range finding, improving physical fitness, and sniper ballistics, in particular the effect of British and German bullets on trench defences. Hesketh-Prichard found that the officers and NCOs who attended his courses 'were extraordinarily keen. They liked sniping and still more obser-vation because they felt that here, at last, in the great impersonal war, was an opportunity for individual skill.'[1]

In between courses the instructors would go back into the line to maintain their battlefield experience and keep abreast of new tactical developments. While making tours of the trenches – when sniping was still something of a novelty – the instructors found themselves looked upon in 'the light of performing animals who would give some show of greater or lesser interest'.[2] Eventually deputations were made to higher authority to allow the instructors to get on with their work, instead of carrying out feats of marksmanship for curious onlookers.

The trainers were agreed that successful sniper recruits needed to be above average infantrymen who were reliable, intelligent, good shots, physically fit, and educated to a basic standard. Inevitably it was difficult to find first-rate material as good men were jealously guarded by their battalion and company commanders, but if these officers could be persuaded that sniper training would improve unit performance then a supply of good recruits was usually forthcoming. The commandant of the Second Army SOS School, Major N. A. D. Armstrong, recommended that snipers be drawn from men who had led an outdoor life, preferably, 'game hunters, trappers, prospectors, surveyors, lum-berjacks, poachers etc.'[3] – although such men would have been hard to find in a predominantly city-recruited British Army.

As the war progressed, snipers were removed from oper-ations at company level and assigned to the battalion, to live and work alongside the battalion HQ. This development had sound tactical advantages in allowing the snipers greater freedom of movement and action. In the First World War – as in later conflicts – snipers assigned directly to companies and platoons found themselves caught up too closely in basic infantry fighting and consequently were unable to achieve the potential that a higher level of control allowed.

The sniper establishment varied between battalions but a fairly standard figure was sixteen men and two NCOs under the supervision of a sniper or intelligence officer. The men would operate in eight two-man pairs, alternating between the roles of sniper and observer. Major F. M. Crum recommended that the sniper section be spared any unnecessary fatigues and 'get most of their nights in bed' but he warned that the senior NCO should be 'a good disciplinarian, so as to supervise the care of valuable instruments and prevent the tendency of all "employed men" to deteriorate into scallywags'.[4]

SNIPER VERSUS SNIPER

Hesketh-Prichard divided the sniping war on the Western Front into four phases. In the first, the Germans dominated the field, roughly the period 1914–15 when British inexperience cost them heavy casualties. In the second, 1915–16, the Allies fought for mastery of the front line. In the third, late 1916 to early 1918, the Germans became far more cautious and cunning stratagems had to be adopted to get them into the open. In the last phase, in the more fluid warfare that followed the German March Offensive of 1918, there were many opportunities for snipers to move out of the trenches and engage in scouting and sniping in the open.[5]

For the keen sniper the second of these phases was arguably the most satisfying, as Hesketh-Prichard later wrote: 'The first work to be done in the trenches was the organized annihilation of these skilled German snipers, and I think that this was easier in that they had it their own way for so long.'[6] Working on the principle that it takes one to catch one, snipers proved most successful in counter-sniper operations. Once an enemy sniper position was located, a trap would be set to lure him out from cover, but as British skills improved so the Germans grew more careful and locating the sniper demanded ever more ingenious methods.

Leaders in the field of camouflage, the French had been mass-producing dummy heads for decoy purposes from 1915 onwards. British snipers visited the great camouflage works at Amiens and began to order up dummy heads of British soldiers made out of papier-mâché. Painted by artists the dummy heads could only be distinguished from the real thing at very close ranges.

In an area where a German sniper was known to be operating, a dummy head was fixed on to a stick and then carefully raised over the parapet by the counter-sniper team. One man held the head in position, while the other stood behind him holding a periscope. When the dummy head was struck by the sniper's bullet it was lowered as realistically as possible to feign death, then the periscope was positioned exactly at the same height and along the line of the entry–exit hole made by the sniper's bullet. The source of the shot could then be located, and sometimes the sniper could be seen disappearing behind cover.

To improve realism, a lighted cigarette might be placed in the dummy head's mouth and smoked through a length of rubber tubing. Hesketh-Prichard laconically recalled: 'It is a curious sensation to have the head through which you are smoking a cigarette suddenly shot with a Mauser bullet, but it is one that several snipers have experienced.'[7]

Some snipers – notably the iconoclastic McBride – considered papier-mâché dummies to be unnecessarily cumbersome, preferring instead to use a practised eye to pinpoint the enemy. Others, like Hesketh-Prichard and Armstrong, were enthusiastic proponents. In the First Army's area of operations, for example, in one trial period, a total of 67 out of 71 enemy snipers were located using dummy heads.

Once the enemy sniper's position had been found, the next stage was to eliminate him. Cunning was needed to deal with an experienced sniper, as Private Raymond Hall of the Manchester Regiment discovered during a period in the trenches in March

1917, around the once notorious Hohenzollern Redoubt. Although by this time a quiet sector, Hall's battalion had suffered seventeen fatalities in the space of two weeks from a lone German sniper.

Hall worked out the position of the sniper but rather than take him on frontally, decided to crawl out at dusk through No Man's Land into the maze of half-destroyed trenches that remained of the Hohenzollern Redoubt. There he hoped to fire an oblique shot into the well-concealed enemy position. Hall instructed two of his comrades to fire towards the German lines in order to attract the sniper's attention. Although Hall could not see his quarry, the muzzle flash from his rifle was clearly visible in the fading light. One carefully aimed shot, four feet back from the flash, dealt with the sniper, who was later seen being dragged back through the enemy trenches by two of his own men.[8] The German had grown fatally careless with easy success and when up against an intelligent sniper had paid the price. Hall noted that from then on sniping ceased on that sector of the front.

As the war continued there was a steady improvement in sniping equipment and tactics. More emphasis was placed on securing good-quality ammunition; once a sniper had discovered a batch of ammunition that gave consistent results then he would acquire as large a quantity as possible for his own use. While the Germans had been using armour-piercing (AP) ammunition since 1915, it was not until later in the war that British snipers received a regular supply. Initially AP ammunition had been used against steel loophole plates but subsequently it was used increasingly to knock out enemy machine-guns. One well-placed shot against a machine-gun's breech mechanism would put it out of action; more efficient than attempting to pick off the gun crew which might easily number six men or more.

In order to counter the increased penetration of AP ammunition, double loophole plates were introduced, first by the Germans and then by the Allies. The second loophole plate was positioned behind the first on a sliding groove. Not only did this provide greater overall protection but as the sniper fired through the two loopholes at an angle it made it far harder for a counter-sniper to get his bullet through the two offset holes. The double loophole plate was, however, much heavier and more cumbersome than the single plate, and was used in fixed sites rather than carried around from position to position.

The use of sniper teams – a sniper and an observer equipped with binoculars and/or telescope – became steadily more common as the war progressed. Some snipers maintained that the 'lone wolf' approach was more effective but the majority preferred to work in teams, and by 1916 the sniper schools had made the two-man team the official sniping unit.

There were many advantages for snipers operating in pairs. Looking through binoculars and especially a telescope is very stressful on the eye and within twenty or thirty minutes the quality of vision deteriorates, but by swapping duties the team was able to maintain effective observation for long periods of time. In addition, by using binoculars (typically around 6-power) and telescopes (ranging from 20- to 35-power) in combination, the likelihood of spotting a target was greatly increased: while the man with the binoculars scanned the general area his tele-scope-equipped companion concentrated on specific points.

A skilled observer could act as a shooting coach, verbally guiding the sniper on to his target to ensure a hit otherwise impossible with a relatively low-powered rifle-mounted tele-scopic sight. The observer was also in a much better position to tell if the target had been hit, his view not being obscured by the rifle's recoil and muzzle blast. On the subject of recording hits the American sniper, H. W. McBride, wrote: 'During our early days at sniping I was greatly amused at the tremendous "bag" made by our various lone-wolf snipers who worked up and down the trenches. These chaps *never* missed a shot, to hear them tell about it.'[9]

On a psychological level the paired team worked at an advantage. The monotony of spending hour after hour in a cramped position was greatly relieved if the sniper had a com-panion, and he was more likely to stay alert if not alone. Also, the observer was able to provide armed protection to the sniper when close to the enemy – the knowledge of which was a morale boost to the often hard-pressed sniper.

The optimum range for sniping varied according to terrain, differing national approaches and individual preference. Because of the relatively short distances from trench to trench, ranges were often as little as one or two hundred yards. But as front-line targets were usually very small – the top of a helmet or any eye looking through a loophole only two inches in diameter – a fine degree of accuracy was required. In the British sniper schools, men were trained to ensure a head shot at four hundred yards; the Germans preferred to shoot at shorter ranges, taking up positions in No Man's Land if necessary. McBride considered targets up to six or seven hundred yards away as worth taking on. Further than that, he claimed, sniping was largely unproductive, although in ideal conditions an excellent shot might hit a fully visible man at a range of up to a thousand yards.[10]

There were three firing zones where the sniper was able to take up his position: in the front line, in No Man's Land or behind the front line. The front line was the main area of sniper activity, the place where a sniper could adopt a concealed and protected position that was closest to the enemy trenches. Loopholes dug amid the jumble of wire and sandbags of a parapet had the advantage of making it difficult for the enemy to locate the origin of a shot, but front-line positions had a number of disadvantages. The chief of these was the restricted field of fire, both laterally (confined to a few yards either side of the loophole) and in depth (a particular problem for the Allies whose forward trenches were usually sited at a lower level than the German front line). And, over time, even the best-concealed front-line position would become known to the enemy, who would either adopt appropriate caution or set about destroying it.

No Man's Land was the preserve of what Hesketh-Prichard called the 'wild boys'. Although enterprising soldiers (such as Private Hall, noted above) were able to crawl into good firing positions, operating in No Man's Land was fraught with danger and was rarely carried out by experienced snipers. Apart from its nerve-sapping qualities, sniping between the lines was rarely that effective. The field of fire was even more limited than that found in the front line; movement was virtually impossible in daylight against an observant enemy; and only one shot (in exceptional circumstances, two) could be made from such an exposed position.

One such 'wild boy' was Ernst Jünger, a German soldier who, along with Erwin Rommel, was one of only thirteen infantry officers to receive the coveted *Pour le Mérite* during the entire war. In the summer of 1918, Jünger's unit was holding a position around Rossignol Wood which had been particularly troubled by rifle grenades fired from a forward sap in the British front line. Enlisting the help of his batman H., Jünger set out to deal with this post. Jünger's account not only describes the action but gives an idea of the intensity of experience that made sniping in No Man's Land such a fearful yet exhilarating activity. Slipping through their own wire into No Man's Land, Jünger and H. slowly crawled through the long summer grass that gave them cover from enemy observers. Jünger wrote:

I made long pauses and then crept forward again, always careful not to move a stalk of grass from its place. The muddy ground was cracked, and so hot that it almost burnt our hands. In this manner we got along by degrees and at last reached the crest of the rising ground. There we found a recent shell hole that must have been the result of a shell with a quick-action fuse, for it was as flat as a stork's nest and all the grass around it was burnt black and close like a dirty blanket.

I slowly propelled myself into this shell hole and brought my eyes on a level with its rim. This done I had a view. I saw little but all I wanted. Beyond, on the level ground, the enemy trenches stretched without sign of life, their outlines dancing in the heat that flickered over the surface of the ground. There was only one gap, perhaps two metres wide, that gave a glimpse into the trench. This was all I could observe; it did not seem to promise much, and it might be all there would be till night fell. But when one lies in wait for the most dangerous being in the world, one must spare neither time nor pains. So I decided to wait and not to move my eyes once from that spot.

Suddenly a sound rang out – a sound foreign to this noontide scene, an ominous clinking as of a helmet or a bayonet striking the side of a trench. At the same moment I felt a hand grip my leg and heard a low-breathed whistle behind me. It was H., for he had passed those hours in the same alert attention as I.

I pushed back with my foot to warn him, and at the same

moment a greenish-yellow shadow flitted across the exposed part of the trench. It was a tall figure in clay-coloured uniform, with a flat helmet set well down over his forehead and both hands grasping his rifle, which was slung from his neck by a strap. It must have been the relief as he came from the rear; and now it could only be a matter of seconds till the man he relieved passed across the same spot. I sighted my rifle on it sharply.

A murmuring of voices arose from behind the screen of grass, broken now and again by suppressed laughter or a soft clanking. Then a tiny puff of smoke ascended – the moment had come when the returning post lit a pipe or cigarette for the way back. And in fact he appeared a moment later, first his helmet only, next his whole figure. His luck was against him, for just as he came in the line of aim, he turned round and took his cigarette from his mouth – probably to add a word that occurred to him during the few steps he had come. It was his last, for at that moment the iron chain between shoulder, hand and butt was drawn tight and the patch pocket on the left side of his tunic was taken as clearly on the foresight as though it were on the very muzzle of the rifle. Thus the shot took the words from his mouth. I saw him fall, and having seen many fall before this, I knew he would never get up again. He fell first against the side of the trench and then collapsed into a heap that obeyed the force of life no longer but only the force of gravity.[11]

Jünger and H. crawled back to their own trenches, reaching safety just in time as British machine-gun fire raked No Man's Land. Despite success, Jünger's account reveals some of the limitations of sniping in No Man's Land, not least the difficulty of movement and the restricted field of fire. Against a better prepared unit the German snipers may not have returned safely to their own line.

Most veteran snipers preferred to operate a short distance back from the front line, the loss of 100–200 yards in range set against the gain of a greater field of fire and the ability to pick better sites to command the opposing trenches. Rather than construct and operate from a single position, snipers were encouraged to build a series of posts. They would then move from one to the other, firing only a few shots from each, thus making it exceedingly difficult for the enemy to locate them.

If a position was discovered the conscientious sniper rendered it inoperable, and moved on to build a new one. The siting of a sniper post demanded great care, as the enemy should be unaware of its construction and its siting should not make it seem a 'likely' position. A ruined building occupying a commanding feature would merely act as a beacon for enemy artillery fire; it was much safer, for example, to dig a camouflaged position secretly in an open field.

Movement into exposed sniper posts had to be hidden from the enemy, either by careful stalking or under the cover of darkness. Failure to do so was a common cause of casualties among inexperienced men. One of Captain J. C. Dunn's respondents left a terse note of this danger during the latter part of the 1918 German offensive: 'Our snipers did not get into their hides until it was light, so they were seen; already half of them have been hit.'[12]

A constant problem for snipers was preventing other troops from using their positions, especially loopholes in the front line. Notable offenders were curious visitors to the trenches, sentries, and artillery Forward Observation Officers, who found that sniper posts made excellent observation platforms for the registration and correction of their guns. As these unwanted guests tended not to follow correct camouflage procedure they regularly gave away the position to the enemy. Especially important was the adoption of measures to keep the actual loophole properly 'gagged' with a piece of rag, and to ensure that a heavy cloth curtain was in position behind the hole to prevent give-away light showing through when the gag was removed. In exasperation, Major Armstrong complained that 'it is most difficult to maintain exclusiveness of loopholes in the front line' and called for a formal battalion order to keep outsiders away.[13]

Once in the post the sniper or sniper team would produce range cards, simple hand-drawn maps with the ranges of likely targets drawn in. Consequently, when a target presented itself there would be no delay in estimating ranges: the observer would rattle off the position and range, the sniper swiftly adjust his telescopic sight and fire. It was important that each sniper should keep an accurate 'score book', a log listing the sight settings, ammunition, wind, temperature and other weather conditions. From these details the sniper was able to build up an overall picture of sniping conditions and how they affected the overall accuracy of his shooting.

Listing the number of kills made was something of a problem area – especially for the inexperienced and boastful – and veterans were often regaled with extraordinary stories of sniping prowess from men with little knowledge or ability in the discipline. Snipers of the calibre of Hesketh-Prichard, Armstrong or McBride preferred not to make lists of their shooting achievements. McBride wrote: 'You might even mark down the "bulls" as you make them – but this will often be a very uncertain matter and you will have to let your conscience be your guide many, many times.'[14]

CAMOUFLAGE AND OBSERVATION

By mid-1916 camouflage was taken very seriously by the combatant nations and camouflaged objects were being produced on an industrial scale. On the British section of the Western Front, camouflage was the responsibility of the Special Works Park. Borrowing much from the French, their main function was to develop and manufacture camouflage to cut down the visibility of guns and large pieces of equipment with garnished nets, and to construct artificial observation posts.

Help was also given to snipers in the form of dummy heads for counter-sniper work and camouflage clothing made out of painted canvas and/or scrim. The early pattern clothing was based on an oversize greatcoat made from canvas with a large hood. The Germans had been the first to use these camouflage robes, and found them useful when crawling in No Man's Land. Eventually, however, they were considered to be overly cumbersome and boiler-suit style garments were subsequently worn, with a detachable hood, rifle cover and gloves. Using these suits, along with local vegetation applied around the garment, a stationary sniper could remain undetected to a distance of only a few feet. These sniper robes were a forerunner of the modern sniper's ghillie suit. When full body cover was unnecessary a camouflaged hood was issued; completely covering the head it had a small slit for the eyes, over which was attached a gauze eyepiece, thereby making the sniper's face almost invisible.[15]

The camouflage works specialized in producing elaborate artificial observation posts which included false trees, made from bark-covered steel, hollowed out milestones and even 'dead horses' constructed with wire-framed canvas. These were occasionally employed by snipers but were considered too static for regular use; after only a few shots these over-elaborate positions would be given away. And few snipers could have relished being encased in such structures, preferring more modest positions with easier and safer means of entry and exit.

Through his long experience of watching the enemy's every movement, the sniper inevitably developed into an expert observer, so that in due course he became a valuable part of the intelligence-gathering net. Up until 1916–17, however, information gathered in by snipers had rarely been acted upon, but as sniping became more organized, they were instructed to conduct systematic observation, the results of which were routinely sent back to battalion and brigade headquarters.

Equipped with binoculars or a 20-power telescope the sniper had the time and the observational skills to notice troop movements, construction of new trenches or the deployment of enemy artillery. A good observer could spot the rotation of units in the front-line trenches; his telescope able to pick up cap badges or national and state cockades at ranges of up to one hundred and fifty yards.[16] Sniper officers considered that intelligence-gathering trench raids, which were regularly a cause of heavy casualties and yet something of a mania among certain units, were often unnecessary; they could gain the necessary information simply by looking through a telescope.

Actual shooting played a relatively small part in the sniper's time, especially when the enemy was well dug-in on a settled section of the front. In these conditions observation became the sniper's main task. Lieutenant F. P. J. Glover had graduated from the First Army SOS School to become sniper officer in the 1st Battalion, the East Surrey Regiment. As part of his duties, Glover wrote daily summaries of activity observed by his men. His summary of snipers' reports for 25 January 1917 is fairly typical:

Enemy Movement

Considerable enemy movement was noted throughout the day in the enemy trenches in the neighbourhood of A10c 2.2 [a trench junction] as follows:

7.10 a.m.   Three men were seen walking over the top behind the front line, moving to the right, and disappeared into the trench. They all wore soft caps and greatcoats and one was carrying a sack on his back.

11.10 a.m.   A German sniper, axe in hand, got on to the parapet of their second-line trench and, crouching down, proceeded to chop away at something. Our sniper fired and the man dropped, rolling over on to the trench. He was wearing a round soft cap, short tunic and top boots, his general appearance being very dirty. Some three minutes later a periscope was raised near this spot and was moved about for over an hour.

2.45 p.m.   A man wearing a green-coloured cap, with a shiny peak, and a green greatcoat looked over the second-line trench. He glanced at our line, disappeared, reappeared more to the left and then apparently moved off to the left.

3.10 p.m.   A passing glimpse was obtained of a Hun passing down the front line from left to right and again of another ten minutes later.

New timber is also visible between the first and second lines (A10c 15.15), resembling in appearance our mining timber, while about A10m a quantity of vapour was seen rising from the enemy front-line trench.[17]

The death of the enemy sniper apart, these summaries made for prosaic reading, but to intelligence officers higher up in the chain of command such snippets of information could be invaluable, for example, the general increase of enemy activity, the significance (or otherwise) of the green-uniformed soldier; the bringing up of fresh timber. In themselves they amounted to little, yet combined with other intelligence they might well help confirm preparations being made for a forthcoming attack or merely the turn-around of front-line units.

The British were fortunate in being able to call upon the skills of the Lovat Scouts, for the most part Highland gamekeepers (ghillies) raised on the Scottish estates of Lord Lovat. Two battalions had been raised on the outbreak of war in 1914 and had seen service at Gallipoli and in the Middle East. In 1916 a unit of Lovat Scouts (Sharpshooters) was sent to the Western Front. The original intention was to employ them as snipers but their extraordinary ability as 'glassmen' ensured that they were kept on observational duties.

Divided into nine groups, each about twenty men strong, the Lovat Scouts were attached to the Army at corps level. They were then made available to front-line commanders for reconnaissance and other observational duties. Their deer-stalking experience in the Highland glens, where they had been trained to use powerful telescopes from an early age, gave them an unrivalled edge in locating German positions and troop movements, not just in and around the front-line trenches but also deep within enemy territory. Hesketh-Prichard worked closely with them and formed a deep admiration: 'The Lovat Scouts never let one down. If they reported a thing, the thing was as they reported it.'[18]

In offensive operations the sniper's role changed from that of a simple defender of his trench lines. This became evident in the Allied battles of 1917 and even more so in the fighting that followed the great German offensives of 1918. Immediately before an attack the sniper would go ahead into No Man's Land and take up a position where he could harass the enemy; his main target would be machine-guns which otherwise could pin down an advance for long periods. To this end, snipers were liberally equipped with armour-piercing bullets. A Canadian citation demonstrates the sniper's usefulness in the attack: 'In the battle east of Arras on 26 August 1918, the advance of the 4th CMR Battalion was held up by strong enemy machine-gun fire. Three snipers went out to the flank and picked off the crews of four machine-guns. They then rushed the post, capturing four guns and fourteen prisoners.'[19]

Fighting out in front of the main body of troops the sniper had to travel light, and be flexible enough to fit into any environment. Sniping in the trenches – with its paraphernalia of steel loophole plates, sniperscopes and artificial camouflage – had necessarily led to a static mode of operation which left certain men at a disadvantage in the open battles of 1918. As the Germans slowly retreated, their own snipers performed very effectively in a defensive role, particularly as the advancing Allied troops now presented more easy targets. An Allied report noted that, 'at Le Quesnel on 9 August [1918], one Boche sniper did more damage than four machine-guns which were firing from the same locality. The guns were located, the sniper was not.'[20]

As the Allies went over to the offensive in the summer of 1918 they were joined by substantial forces from the newly arrived US Army and Marine Corps. Indeed, it was the deployment of the US armed forces that tipped the numerical scales decisively in the Allies' favour. American soldiers had higher shooting standards than those of the other combatants on the Western Front, in part through an enthusiastic hunting tradition in the States and because by 1918 musketry standards among Britain, France and Germany had inevitably declined through the exigencies of war. (It can be argued that the sniper's importance increased further during the latter stages of the war as a result of an overall fall of shooting ability among the rank and file.)

The exploits of Sergeant Alvin York have passed into legend, but a high standard of marksmanship was not unusual among American infantrymen. US Marines were expected to be able to fire well-aimed accurate shots out to eight hundred yards, and one contemporary writer made this comment on the damage caused to German morale from American rifle fire: '. . . aimed, sustained rifle fire, that comes from nowhere in particular and picks off men – it brought the war home to the individual [German] and demoralized him. And trained Americans fight best with rifles.'[21]

American snipers were armed with improved examples of the standard US service rifle, the .30 Model 1903 Springfield, fitted with telescopic sights. The Springfield utilized the tried and tested Mauser bolt action and was built around the new .30in-calibre round, which became the American standard for the next sixty years and more. Its relatively short length made it a well-balanced rifle that was easy to use. An accurate weapon, the Springfield remained in US service as a sniper rifle until the Korean War.

Marksmanship alone does not make a good sniper and US troops received training at the British and Commonwealth sniper schools prior to going into the trenches. Of the overall US performance, American sniper authority Peter R. Senich concludes: 'Unprepared as they were initially, by the summer of 1918 when the total weight of the American Expeditionary Force was finally brought to bear, American snipers, while employed only in limited number, proved to be as effective as their British and Canadian counterparts.'[22]

Whether snipers from one army were superior to those from another is very difficult to assess, and the many claims made for one side against another must lie in the realms of conjecture. British snipers were prepared to concede German superiority between 1914 and mid-1916 but from then on they were sure that the balance had shifted in their favour. This was almost certainly the case in the areas where British snipers were actively working, but it is also clear from contemporary accounts that German snipers continued to give trouble right up until the Armistice.

McBride maintained that the Germans were good snipers at short ranges – 'up to three hundred, possibly four hundred yards' – but were poor shots at greater distances. In agreement with McBride, Hesketh-Prichard made this comment on his opponents: 'Of the Germans as a whole one would say that with certain brilliant exceptions, they were quite sound, but rather unenterprising, and that as far as the various tribes were concerned, the Bavarians were better than the Prussians, while some Saxon units were really first-rate.'[23]

Hesketh-Prichard reasoned that Germans operated effectively at close range because many of their best shots were jaeger forest guards who had worked in Germany's great coniferous estates, where a keen eye was needed for shooting deer and wild boar at short distances. As a result the Germans rarely used spotting telescopes, confining their observational range to that provided by binoculars. By contrast the British, Canadians and Americans were used to firing at long ranges, whether in target shooting or hunting.

Of his own side, Hesketh-Prichard considered the English as 'sound, exceedingly unimaginative, and very apt to take the most foolish and useless risks'; the Welsh were 'very good indeed'; the 'Canadians, the Anzacs, and the Scottish regiments were all splendid' and the Americans 'were also fine shots'.[24] A dissenting view on the performance of Welsh snipers is given in an account collated by Captain Dunn of the Royal Welch Fusiliers: 'Our fellows never ran after the job, nor were they much good at it, as extreme patience is not the strong point of the Welsh. We thought the Germans were better at the game; we were obliged always to take advantage of all cover, otherwise they got busy; they scored several hits.'[25]

If the American Civil War marked the origin of the true sniper, then with the First World War came the arrival of the fully-fledged modern sniper. Sniping had previously been a business carried out by a talented if eccentric few, good shots who had a feel for operating over natural terrain. As a result of the First World War, sniping became a recognized part of the military machine: courses were held, pamphlets issued and textbooks written. Instead of being born and bred to the calling, men could now be trained to be snipers.

Since the First World War the essentials of military sniping have changed little. Certainly rifles have improved and tactics been expanded, but a McBride or a Hesketh-Prichard would have little difficulty following a sniper course at Fort Benning, Georgia, or at the School of Infantry, Warminster. The significance of the First World War lay in setting the pattern which sniping would follow in the succeeding years.

CHAPTER SIX

THE SECOND WORLD WAR

OPENING ROUNDS

As German troops crossed into Poland on 1 September 1939 so began the costliest and most bloody war in history. At first the fighting was confined to Europe as Germany waged war against an Allied coalition led by France and Britain. But the war spread inexorably across the globe. Italy's arrival into the conflict on the side of Germany in 1940 brought the Mediterranean and North Africa into the sphere of military operations. In 1941 the transformation into a worldwide conflict was complete, first with the entry of the Soviet Union, and then of Japan and the United States.

While the war as a whole reflected the technological advances of the previous two decades, the sniper's weapons and tactics remained broadly similar to those of the First World War. There was, however, greater freedom of action for the sniper, as a result of a more open and fluid style of fighting. Also, the wider geographical range of the Second World War forced snipers to operate in very different environments, from the ruins of Stalingrad, to the hedgerows of Normandy and the jungles of New Guinea.

Between the two world wars there had been no real development in the theory and practice of sniping. The Spanish Civil War (1936–39) had witnessed widespread if rudimentary sniping. Soviet military advisers to the Spanish government side had been impressed by the potential of sniping and on their return to Russia they encouraged its development, alongside the already extensive civilian rifle-shooting programmes.

The German Blitzkrieg campaigns against Poland in 1939 and the West in 1940 were so swift and fast-moving that there was little time for snipers to show their prowess. In Germany, while sniping remained a recognized part of infantry tactics, no great enthusiasm had been expressed for the discipline. Only when the tide of war began to turn in Russia, as German troops began to take significant casualties from Red Army snipers, did the German Army renew its commitment.

In Britain, sniping had been allowed to atrophy. Writing in 1942, the First World War instructor Lieutenant-Colonel N. A. D. Armstrong provided this comment on the state of sniping during the inter-war years: 'There appeared to be a tendency among Army musketry men to scorn the sniper – they held that sniping was only a "phenomenon" of trench warfare and would not be likely to occur again.' Armstrong continued his lament, noting how telescopic sights had been packed away and that 'the few experts whose services were obtained to train young soldiers in the art of fieldcraft and sniping returned to the outposts of Empire whence they had come and were heard of no more'.[1]

Although Armstrong was exaggerating the situation, sniping remained a low priority for the Army until the retraining programmes got underway after Dunkirk. The doctrine of sniping, as expounded by Great War pioneers like Armstrong and Hesketh-Prichard, had been incorporated into tactical manuals,[2] but as the Army expanded in numbers during the late 1930s little was done to develop sniping actively within each infantry battalion.

British snipers were engaged in Norway and France, although for the most part their operations were conducted on an ad hoc basis. One such sniper, who achieved success in France and Belgium during the German offensive in May 1940, was Edgar Rabbets, a soldier in a Territorial battalion, the 5th Northants. He was a countryman from Boston in Lincolnshire – capable of catching a rabbit with his hands – and as a crack shot it was almost inevitable that he was appointed a company sniper when his unit was sent overseas.

Rabbets was given complete freedom of operation in order to fulfil his role of eliminating enemy snipers and picking off other suitable German targets. By choice he worked without a partner, finding it easier to move across country as a 'lone wolf'. Although it was standard practice for snipers to work in teams, this was never rigorously adhered to. The sniper officer and author, Captain C. Shore, noted: 'There were times when one met a tough, taciturn, hard-bitten case who preferred to work alone; such men travelled farthest and killed best alone, and a wise commander allowed them full rein.'[3] Rabbets was such a sniper.

During the British retreat to the Dunkirk perimeter Rabbets' battalion was involved in holding the line. German snipers had taken up a position in a deserted Belgian village and Rabbets was ordered forward to eliminate them. Rabbets discovered that he was up against a standard sniper-spotter team. The sniper, Rabbets recalled, 'had got himself up in a roof and knocked a few slates away. He'd got a good field of fire if anyone walked into the square; he was roughly in the centre of one side of the square and his mate was in the corner. And they covered the whole square that way, the one effectively protecting the other.'

The German sniper then took a pot-shot at a British officer entering the square. Rabbets continued: 'I happened to find out roughly where the flash had come from and I went into a house opposite. The sniper was hanging out of the roof; I shot him from the bedroom window and he fell forward.' Rabbets then dealt with the observer, who had made the mistake of firing blindly at the British sniper, so giving his position away: 'I was firing deep from out of the bedroom window, and I wasn't exposed to view. He assumed wrongly that I was a lot nearer the window than I was. And he gave himself away, so that was his lot, because I was able to fire straight into the corner of the square where he was.' For these Germans – as for many other inexperienced snipers – failure was literally fatal.

As the Germans advanced confidently through Belgium and France during May 1940 they presented a variety of targets to a good marksman. On one occasion Rabbets was able to knock out two German infantry support guns; the noise of their own guns made the crews oblivious to his accurate rifle fire.[4]

As a general rule Rabbets preferred to use his talents in fieldcraft to get in close to his targets, thereby improving his chances of a first-round kill. Rabbets was an accomplished marksman, and was capable of gaining a head shot up to a range of four hundred yards from a standard .303 Lee-Enfield. Rabbets generally went for the head rather than the torso: 'It was the best place to kill them; it's a nice white target and you know once you've hit him in the head he's dead. It's as good a target as any, especially with the German helmet which just lines it up for you; you hit him just below the helmet in between his eye and his ear.'[5]

Rabbets combined sniping with intelligence gathering. He illustrated this dual role with an incident which took place on the road to Dunkirk:

If I found an [enemy] unit that was obviously moving into a position that was going to cause us some problems, the sooner my unit knew about it the better. One day I went out and found a German military policeman standing at a crossroads; the only reason they stand at a crossroads is to direct a unit into a new position. I wanted to know what he was doing, so I crawled to within 150 yards range. He gave himself away by continually looking up the road he was expecting the unit to come from, and because there was only one direction towards our lines I knew roughly where they were going to. I shot him and then bundled him out of the way so when the enemy got to the crossroads they wouldn't know where they were going. Then I went back to give my unit this intelligence.[6]

BARBAROSSA AND AFTER

The next phase of the sniping war came with the German invasion of the Soviet Union on 22 June 1941; for nearly four years the two nations would be locked in a titanic struggle. The Soviet Red Army had encouraged sniping during the 1930s, their programme receiving further impetus from experience in the Spanish Civil War (1936–39) and the recent Winter War against Finland (1939–40).

In the latter conflict the Soviet forces had received a mauling from the Finns, and only vastly superior resources finally gave the Red Army victory. The Finns had been particularly adroit in using their hunters as snipers; they had a long shooting tradition, and many of their best shots were eider-duck hunters who were experienced in killing their prey with the greatest accuracy, to minimize damage to the bird's plumage. These men found the slow-moving columns of Soviet troops blundering through the Karelian forests easy game. Simo Häyhä was a farmer and hunter before the outbreak of war and his farmhouse was full of trophies for marksmanship. When the Red Army invaded, Häyhä picked up his rifle and went out 'to hunt Russians', which he did with a vengeance. Although some doubt remains over his final tally, it was claimed he killed more than five hundred Soviet troops before being seriously wounded.[7] The lessons of the Winter War were not lost on the Red Army.

The Soviet sniper rifle was the 7.62mm Model 1891/30 Mosin-Nagant, the ordinary bolt-action Soviet infantry rifle, accurized for sniping and fitted with a telescopic sight. A rugged weapon able to withstand the rigours of the Russian climate, the M91/30 was popular with both Soviet and German snipers, the latter using large numbers of captured examples in preference to their own weapons. The M91/30 remained in frontline service throughout the war, and was used by Soviet-sponsored guerrilla movements right up until the conflict in Vietnam,

The Soviet Union had benefited from the purchase of the Zeiss optical company in the 1930s, and produced the 4-power PE telescopic sight in large quantities (54,000 telescopic-sighted M91/30s were built between 1932 and 1938). After the start of the war a lighter 3.5-power PU scope came into service for use alongside the PE Model.[8]

As a means of upgrading the infantryman's firepower the Red Army took delivery of two semi-automatic rifles, the Tokarev Models 1938 and 1940. Both rifles were manufactured to receive a telescopic mount and scope, and some were issued to snipers. They proved less than successful, however, because of problems in the ammunition feed and the unreliability of the gas system during cold conditions. They were withdrawn from service in the period 1942–43, although, ironically, the Tokarevs influenced

German rifle manufacturers towards the design of semi-automatic weapons.

During the war the Soviet military authorities promoted sniping throughout the Red Army. Their definition of the sniper's role was broader than that of the West, including general sharp-shooting as well as dedicated sniping. Yet the Soviet military authorities made sniping a fully integrated part of infantry tactics, whereas in the West interest in sniping blew hot and cold, according to wartime expediency.

Soviet snipers usually worked in pairs, and were allowed considerable freedom of action. They operated at a low tactical level, assigned directly to companies and platoons. There were normally large numbers of snipers available to junior infantry commanders who were used to handling them in everyday tactical situations.

During 1941–42 the Germans attacked and the Red Army retreated. Consequently most Soviet sniping was of a broadly defensive nature, and apart from local counter-attacks the Red Army sniper's main function lay in slowing down the Wehrmacht's seemingly inexorable advance into Russia. Snipers would take up hidden positions forward of their main line of resistance, firing on enemy reconnaissance patrols, officers and the forward artillery observers whose presence invariably signalled the start of a German assault. In a paper prepared by Lieutenant-General G. Morozov, an example was given of the effectiveness of such tactics:

The company commander, Senior Lieutenant Stepanov, antici-pating a local enemy attack, sent three pairs of snipers to take up positions on the nearest approaches to the Soviet defence lines. One pair, who established themselves near a little bridge across a stream, was instructed to keep a definite zone under observation, paying special attention to the exit from a certain thicket. An automatic rifleman accompanied this pair of snipers. They were instructed that when the enemy's vanguard reached a specific line they should withdraw to the main position. While withdrawing, they were to halt at intermediate points and open fire. Similar precise instructions were given to the second pair, who advanced to the mouth of the gully, and to the third pair, who took up a position at the bend of a stream.

After some time enemy scouts appeared. Adapting themselves to the terrain, they advanced toward the company's defence line. The snipers waited until the scouts had emerged from a corn field and then opened fire. Most of the scouts, including an officer, fell.

The enemy then brought up machine-guns, which the snipers held in check for half an hour. The Germans apparently decided that the line where the snipers were stationed was the advanced defence line, and opened artillery fire upon it. By then, however, the snipers had withdrawn, and the German guns shelled in vain.

As soon as they reached the main Soviet positions on the advanced defence line, the snipers became part of the general fire system, definite targets being assigned to them. The enemy did not succeed in advancing beyond the stream. The Soviet battalion, which later launched a counter-attack, repelled the Germans.[9]

Whether such a sophisticated defensive scheme was common – and in this instance so apparently successful – remains conjectural, although the ruse of confusing the enemy as to the real defence line was successfully used by the Germans against the Allies in Normandy and Italy.

While the Red Army tried to use its numerous snipers in as many ways as possible, they were most effective in static set-piece battles. The great struggle for Stalingrad was perhaps the best instance of the war between the snipers, as marksmen became a key element in the battle for possession of Stalin's city on the Volga. In September 1942 the German Sixth and Fourth Panzer Armies were forcing the hard-pressed Soviet troops back into the city. Repeated artillery and aerial bombardments had turned most of Stalingrad into rubble – ideal terrain for snipers.

Snipers were sometimes isolated from their comrades for days on end, even though they might be only a hundred yards away. During daylight a sniper in an exposed position was forced to remain motionless; movement was permitted only at night.

A slightly cut-down version of the Mauser Gewehr 98, the 7.92mm 98k was used extensively by German snipers during the Second World War.

The German 7.92mm Gewehr 43 was a semi-automatic rifle manufactured with a sniping capability, although most snipers preferred the simpler and more accurate bolt-action 98k.

The 7.62mm M1891/30 Mosin–Nagant was the standard rifle used for sniping by the Soviet Red Army during the Second World War.

A Japanese 6.5mm Type 97 sniper rifle. The case for the telescopic sight is shown above the rifle, and the folding monopod is attached to the front band of the stock.

A US .30–06 Springfield M1903A4 sniper rifle. In contrast to the M1903A1, this rifle was not fitted with iron sights and relied solely on its 3-power Weaver telescopic sights.

The Israeli 7.62mm Galil sniper rifle has a fold-away stock, shown here in its folded position. Other features include the side-rail telescopic mount and the bipod positioned alongside the forward hand-guards.

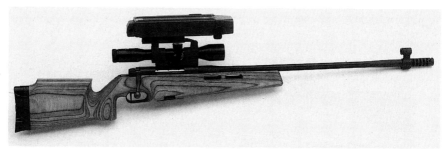

A distinctive feature of the 7.62mm Mauser M86SR, is the wooden stock with ventilation holes to help heat dissipation from the barrel. This example is fitted with a laser range finder.

A Heckler and Koch MSG90 sniper rifle. This is the military version of the PSG-1, and is shown here with adjustable stock and bipod.

The Walther WA 2000 is a revolutionary sniper rifle featuring a bull-pup configuration, in which the magazine is situated behind the trigger assembly.

This Austrian Steyr SSG 69 sniper rifle has been fitted with a suppressor for covert operations. Spacers are visible at the butt end of the stock.

The Beretta sniper rifle has a finely made wooden stock, with thumb hole, adjustable bipod and a flash hider attached to the muzzle of the barrel.

The Sako M86 is manufactured in Finland, and along with its wooden stock it is fitted with a muzzle brake and flash hider.

A Yugoslav M76 sniper rifle in 7.92mm calibre, fitted with a night sight. The M76 is similar in overall design to the Soviet SVD.

The Swiss SIG SSG series of sniper rifles employ the finest materials and workmanship, and are accordingly well regarded.

A British Parker Hale M82 sniper rifle, shown here with a Pilkington IWS night sight. The M82 was bought by Canada and Australia as the basis for their sniper rifles.

The British 7.62mm L96A1 sniper rifle, the standard infantry model of Accuracy International's sniper rifle system. The high-impact plastic stock is fitted over an aluminium frame to give the rifle its distinctive outline. (*Accuracy International*)

The .338 (8.6mm) Super Magnum rifle is designed for long-range sniping, and is fitted with a Bausch & Lomb 10-power telescopic sight. (*Accuracy International*)

Accuracy International's Covert model is a 'take-down' rifle which can be fitted into a suitcase. The extra 'thick' suppressed barrel is clearly visible here. (*Accuracy International*)

Accuracy International's AW series is replacing the PM system; this is the AWP model, designed for police and counter-terrorist use, and fitted with a variable-power 3-12 × 50 Schmidt and Bender scope (*Accuracy International*).

The sporting origin of the Remington Model 700 rifle is apparent, depicted here in its 7.62mm M40 version with a variable-power 3-9 Redfield telescopic sight.

The M40A1 is a US Marine development of the M40, and utilizes a synthetic stock with the Remington action. A 10-power Unertl sight is fitted as standard.

The US Army M21 sniper rifle was a redesignation of the accurized 7.62mm M14 rifle used in Vietnam. Unlike most other sniper rifles the M21 employs a semi-automatic action.

The .50 (12.7mm) Barrett M82A1 is an ideal weapon for firing at targets in excess of 1000 metres; its heavy bullet can pierce light armour and destroy military equipment at these ranges.

(Unless otherwise specified, all pictures courtesy of Ian V. Hogg)

Snipers faced all the usual discomforts of military life in the front line, exacerbated by the specialist nature of their work. They were invariably hungry and thirsty, subject to bursts of intense fear and long stretches of boredom, and in extreme cases forced to urinate and defecate where they lay rather than give away a position by movement.

The top snipers gained unprecedented prominence during the Stalingrad battle. Undoubtedly the most famous was Vasili Zaitsev, a hunter from the Urals and already an experienced sniper before arriving in Stalingrad. Between August and the end of October he was credited with killing over a hundred Germans. In addition, he set up an unofficial sniper school in the remains of the vast Lazur chemical plant. The training was rudimentary but sufficient to improve a soldier's chances of survival and his ability to kill Germans – the bottom line of this academy in the rubble. After a course lasting only two days the 'graduate' would immediately take up his position at the front.

The Red Army snipers were a constant thorn in the side of the German Sixth Army command; morale fell and attacks lacked determination when every soldier felt himself under the stare of a telescopic sight. As the Germans became aware of the celebrity of Zaitsev through captured Soviet newspaper reports, they decided that he must be eliminated as part of a campaign to wrest sniper supremacy from the Red Army. Major Konings, the chief instructor of the Zossen sniper school near Berlin, was despatched to Stalingrad. A noted marksman, Konings was considered the best man to take on the wily Zaitsev. Amid the noise and horror of the industrial war machine, the struggle was reduced to its essence – the confrontation of man against man.

Zaitsev and his fellow snipers became aware of Konings' presence, and every effort was made to locate him, as Zaitsev revealed in a postwar memoir:

> Every sniper put forward his speculations and guesses arising from his day's observation of the enemy forward positions. All sorts of different proposals and baits were discussed. I knew the style of the Nazi snipers by their fire and camouflage and without any difficulty could tell the experienced snipers from the novices, the cowards from the stubborn, experienced

enemies. But the character of the Head of the School [Konings] was still a mystery to me. He presumably altered his position frequently and was looking as carefully for me as I was for him.

Then something happened: my friend Morozov was killed and Sheykin wounded – by a rifle with telescopic sights. Morozov and Sheykin were considered experienced snipers; they often emerged victorious from the most difficult skirmishes with the enemy. Now there was no doubt. They had come up against the Nazi 'super-sniper' I was looking for.[10]

Taking along his spotter, Nikolai Kulikov, Zaitsev roamed across the city for several days, cautiously inspecting the enemy front lines for any sign of the German. As the light was beginning to fade one evening they found him. Konings had set up an ambush designed to force a Soviet sniper to give away his position, but Zaitsev had spotted the ruse and now began to close in to locate Konings' exact position. Zaitsev's discovery attracted interest and a political commissar went along to help. The commissar's curiosity almost proved his undoing when, for only a second, he raised himself above the parapet. It was enough for Konings, and his well-aimed shot flung the commissar back into the trench, wounded. Zaitsev explained the final stages of his duel with Konings:

> For a long time I examined the enemy positions, but could not detect his hiding-place. From the speed with which he had fired I came to the conclusion that the sniper was somewhere directly ahead of us. I continued to watch. To the left was a tank, out of action, and on the right was a pillbox. Between the tank and the pillbox, on a stretch of level ground, lay a sheet of iron and a small pile of broken bricks. It had been lying there a long time and we had grown accustomed to it being there. I put myself in the enemy's position and thought – where better for a sniper? One had only to make a firing slit under the sheet of metal, and then creep up to it during the night.
>
> Yes, he was certainly there, under the sheet of metal in no man's land. I thought I would make sure. I put a mitten on the end of a small plank and raised it. The Nazi fell for it. I carefully let the plank down in the same position as I had raised it and

examined the bullet-hole. It had gone straight through from the front; that meant that the Nazi was under the sheet of metal.

Now came the question of luring even a part of his head into my sights. It was useless trying to do this straight away. Time was needed. But I had been able to study the German's temperament. He was not going to leave the successful position he had found. We were therefore going to have to change our position.

We worked by night and were in position by dawn. The sun rose. Kulikov took a blind shot; we had to rouse the sniper's curiosity. We had decided to spend the morning waiting, as we might have been given away by the sun on our telescopic sights. After lunch our rifles were in the shade and the sun was shining directly on to the German's position. At the edge of the sheet of metal something was glittering: an odd bit of glass or telescopic sights? Kulikov carefully – as only the most experienced can do – began to raise his helmet. The German fired. For a fraction of a second Kulikov rose and screamed. The German believed that he had finally got the Soviet sniper he had been hunting for four days, and half raised his head from beneath the sheet of metal. That was what I had been banking on. I took careful aim. The German's head fell back, and the telescopic sights of his rifle lay motionless, glistening in the sun . . .[11]

The Soviet sniper was expected to play his part in offensive operations. As the Red Army began its streamroller drive from the Russian heartland to the gates of Berlin, so the sniper's role changed from delaying the enemy to that of supporting his own attack. Deployed at company and platoon level, the snipers moved to the flanks to cover their infantry, and to eliminate all targets that would slow down the impending assault. These would change according to circumstances, but for the most part consisted of enemy machine-gun and mortar crews. If these could be silenced then the infantry would have a good chance of success.

Once the attack was underway the Soviet sniper extended his range of targets to include local centres of resistance: the embrasures of pillboxes, enemy support weapons which were previously hidden, and officers and NCOs co-ordinating counter-attacks. If German armoured fighting vehicles came into the action, the sniper was instructed to shoot at any exposed crew members or, failing that, to use AP bullets to knock out the vehicles' slits and vision blocks. Assuming the attack progressed according to plan, the sniper's next task was to deal with troops covering the German retreat, especially officers attempting to keep order and form new points of resistance.

The Red Army sniper was expected to show more initiative than his rank and file comrades. He was informed that while raw German troops were careless and regularly failed to utilize available cover, seasoned troops were exceptionally cautious and the most artful stratagems were required to draw them into the open. In one instance, a Red Army sniper went into action with grenades and a bottle of combustible liquid. He hid them in an exposed area near enemy lines, then covertly withdrew to another well-hidden position before setting fire to the explosives with a single shot. The resulting blast drew the fire of German machine-guns and mortars. Their positions were carefully logged by the sniper, who then set about silencing them, using either his own efforts or by calling on artillery support.[12]

As in most other armies the intelligence function of sniping was well recognized, and reports of enemy activity were sent back down the line as a matter of course. Snipers would some-times accompany small reconnaissance patrols, to provide them with highly mobile and discreet protection if they were spotted by superior enemy forces. When under attack the sniper pair would hold off the enemy for sufficient time to allow the patrol to escape, before withdrawing themselves.

During the four years of fighting on the Eastern Front the Soviet sniper justifiably gained a reputation as an effective battlefield killer. This reputation was, however, undermined by the repeated accounts of sniper exploits published by the Soviet propaganda machine: huge scores of dead 'fascists' were appar-ently amassed by every good communist who happened to find a sniper rifle in his or her hands.[13] The real proof of the skills of Soviet snipers lay in the unsolicited comments from German soldiers, and the influence of the Red Army on Germany's own sniper programme.

The quality and quantity of Soviet sniping came as an unpleasant surprise to the Germans, but typically they set about rectifying their own shortcomings with systematic thoroughness. Sniping schools were increased in number and snipers were officially accorded greater esteem with the award of a special *Scharfschützen* (Sharpshooter) badge. Oval in shape, the badge consisted of an eagle's head with oak leaves and was awarded in three classes: the first degree for 20 confirmed kills, the second degree for 40 kills and the third degree for 60 kills.[14] The badge, however, only came into service in late 1944; few were awarded and fewer still worn, for to be captured in combat with such an emblem was to invite certain and fatal retribution.

German sniping had an influential if unlikely patron in the form of Heinrich Himmler. From the outset of the war the Waffen-SS had taken an interest in sniping, although its efforts were initially hampered by shortages of adequate equipment, having to make do with captured Polish sharpshooter rifles and telescopes and mounts manufactured from Czech sources. As the war progressed Himmler fought hard to encourage sniping within the ranks of the SS. In 1944, while trying to persuade the armaments minister, Albert Speer, to provide him with increased supplies of telescopic-sighted rifles, he explained that an SS sniper with 50 confirmed kills received a wristwatch, 100 kills gained a hunting rifle, while 150 produced an invitation from the *Reichsführer* himself to go hunting for deer and chamois.[15]

The standard German sniping rifle was the 7.92mm karabiner 98k, a slightly shorter version of the bolt-action Gewehr 98 used by the infantry in the First World War. Although an old design the 98k was popular with snipers because it was accurate and reliable. Snipers received the best-quality rifles from the manufacturers and their own unit armourers were able to make minor improvements. A variety of telescopic sights was issued for the 98k, ranging from an inadequate 1.5-power to 4-power and commercially acquired 6-power scopes.[16]

As part of a requirement to increase the ordinary rifleman's firepower, the German Army introduced the Gewehr 43, a self-loading rifle firing the full-power 7.92mm round. Developed from the earlier (and unsuccessful) Gewehr 41 and influenced by the Soviet Tokarev rifle, the G43 had been designed with an integral telescopic sight mount, enabling its easy use as a sniper rifle. Despite the advantage of a faster rate of fire, the G43 was not popular with experienced snipers; they found it wanting in long-range accuracy, besides having a tendency to jam at critical moments. Matthias Hetzenauer, the German Army's top wartime sniper with 345 confirmed kills, was a firm exponent of the 'one shot, one kill' philosophy. He used a 98k by choice, explaining that 'snipers do not need a semi-automatic weapon, if they are correctly used as snipers'.[17]

By 1943–44 German sniper training was conducted at the highest level; experienced snipers were withdrawn from the front line to pass on their knowledge to sniper recruits, themselves selected from the best infantry marksmen. Much had been learned from experience on the Eastern Front and the skills of the 'wily Bolshevik' were incorporated into the lessons. As the trainees were already superior marksmen, particular emphasis was placed on observation, camouflage, general fieldcraft and the need for the sniper constantly to use his imagination to outwit the enemy. One training film stressed the need to 'acquire the primitive feeling for the countryside shown by the hunter'.[18]

Hetzenauer and his two sniper comrades from the 3rd Mountain Division, Sepp Allerberger (257 kills) and Helmut Wirnsberger (64 kills,) agreed that successful snipers must be good marksmen and possess the virtues of patience, perseverance and good judgement. More specifically, Hetzenauer emphasized the necessity of cunning, and how 'the better "tactician of detail" wins in combat against enemy snipers. The exemption from commitment to other duties contributes essentially to the achievement of high scores.' In accord with British and American sniper practice, Hetzenauer recommended that snipers should be chosen from 'people born for individual fighting such as hunters, even poachers, forest rangers, etc.'

In contrast to the Soviet pattern of deployment to companies and platoons, German snipers were organized at battalion level;

they reported directly to the battalion HQ and were assigned to companies for temporary assignments. Wirnsberger was in command of a battalion group of twenty-two snipers, although he explained that in the latter stages of the war – as the numbers of fully trained snipers decreased – 'they were often assigned and given their orders by the division'. Working alongside the snipers were a number of marksmen selected from the line companies and armed with telescopic rifles. They were not trained to the same high standards as the regular snipers and served in their companies as ordinary soldiers, but they could be relied upon to hit man-sized targets at ranges of up to four hundred metres.

By way of comparison, Hetzenauer estimated that he could hit a standing man at between 700–800 metres, while Allerberger and Wirnsberger both gave a figure of 600 metres. German snipers were trained to put out long-range harassing fire, to show the enemy that they were not safe and force them to go to ground. At such ranges – in excess of 700–800 metres – it was difficult to confirm the success of a shot, although Hetzenauer claimed a one-off hit at 1100 metres against a man standing in the open.

German snipers generally worked as pairs in the usual sniper-spotter configuration. Attention was paid to the elimination of the standard priority targets: other snipers, observers, officers and the crews of infantry support weapons such as mortars and machine-guns. On certain occasions, Hetzenauer explained that he took up a concealed position in the enemy front line 'at night before our own attack. When our own artillery opened fire, I had to shoot at enemy commanders and gunners because our own forces would have been too weak in number and ammunition without this support.'

Defensive actions suited experienced snipers as the enemy normally had to cross open fire zones with limited cover; officers were particularly vulnerable, as they were forced to reveal themselves as they urged their troops to push home the attack. During a delaying action, snipers were often used in preference to a machine-gun; they could move around from place to place, laying down accurate fire without giving away their positions.

The British sniper and author, Captain C. Shore, discovered

the effectiveness of German defensive sniping, citing one instance of an entire battalion of the 51st Highland Division being pinned down by a few German paratroop snipers during an engagement in Sicily. Despite heavy shelling the snipers kept up accurate fire at a range of six hundred yards, before eventually retiring with minimal loss, using fieldcraft of the highest order.[19]

During the campaign in north-west Europe, after the Normandy invasion, Shore recounted the following story as an illustration of the quality of German sniping. Two British soldiers were sniping at German troops, using a hedge as concealment. Shore explained that his informant was standing and the other soldier, a corporal, was kneeling down alongside the hedge: 'Suddenly there came a throat-rattling cry from the corporal and his companion, looking down, saw that he had been hit, the bullet having entered the head just below the rear brim of his steel helmet. He knelt beside the NCO and was hastily fumbling for his field dressing when he felt a stunning blow at the back of his neck and head and knew nothing more until he recovered consciousness an hour later in an emergency dressing station.' The German sniper had hit both men, almost in the same spot, and the fact that Shore's informant had been leaning over the NCO had 'resulted in the bullet ploughing up the back of his head [to emerge] through his steel helmet. In the corporal's case the bullet had gone straight into the head and hit the brain.'[20] Shore noted that the range was at least three hundred and fifty yards and to place two shots in rapid succession into men hidden behind a hedge was excellent sniping.

Besides good shooting, German skill in sniping was based on sound observation and detailed attention to camouflage. The latter was an important factor in every soldier's training and for snipers it was taken to greater lengths, with the exhortation to 'camouflage ten times, shoot once'.[21] German snipers were instructed to wear camouflage clothing at all times, and over the course of the war many different camouflage uniforms came into service. The reversible winter (white) and temperate (brown/ green pattern) hooded jacket was popular although there was always the problem of keeping the white from showing when the temperate side of the jacket was worn. Given a wider freedom of choice than the ordinary soldier, snipers adapted uniforms to suit their own particular preferences. Earth and soot were employed to disguise face and hands, although the Germans also used various veils and mosquito nets to cover the head, disguising the face as well as breaking up the outline of the distinctive German helmet.

Whereas the Allies used a combination of artificial camouflage (burlap and canvas sacking, etc.) supplemented by natural camouflage (grass, leaves, straw, etc.) attached directly to the uniform, the Germans concentrated on natural camouflage tied to wire or string nets. For open terrain the Germans constructed camouflage fans, consisting of a half-circle frame, made either of wire or wood and string, covered in natural camouflage. The fan was pushed slowly ahead of the sniper, and if done with sufficient care it provided reasonable concealment while crossing exposed ground. The more advanced contraptions were nick-named umbrellas and could be folded away after use, in the manner of a collapsible umbrella. The sniper Sepp Allerberger used one on the Eastern Front, but others found fans and umbrellas too cumbersome, regarding them as training tools best left behind at the sniper school.

Other forms of personal camouflage included the grass mat, a large grass-stuffed net which completely covered the sniper; in open and static positions it provided excellent concealment but prevented movement. Wire-framed hoods were fitted to the sniper's shoulder, thereby allowing him to turn his head without moving the camouflage – a useful advantage when scanning a broad panoramic field. String nets could be worn over the standard camouflage uniform, and be garnished with natural camouflage.

The sniper also needed to camouflage his weapons and equipment. Rifles were sometimes painted with a disruptive spotted pattern, although it was more common to wrap sacking around the butt and forestock along with strips of natural camouflage. To prevent light from reflecting on optics – both telescopic sights and binoculars – paper covers with horizontal slits were fitted over the front lens; the loss of vision being set against a lesser chance of discovery.

To complement the attention given to camouflage techniques, German snipers were encouraged to use ruses and stratagems to force the enemy to reveal their positions. Perhaps as a result of encountering French and British dummy figures on the Western Front during the First World War, the Germans used them extensively during periods of positional warfare. In order for a dummy to be effective against any but the most green troops it had to be moved in a naturalistic fashion. As a consequence elaborate devices were adopted using pulleys and wire to make the dummy move its head and arms in a 'human' manner. Sometimes rifles were fitted to the dummy and fired using a length of string. The normal practice was for the spotter/observer to operate the dummy while the concealed sniper looked through his sights, ready to fire if the enemy gave himself away by shooting at the decoy.[22]

The quality of training and the combination of initiative and toughness typical of the German infantryman, made the German sniper a deadly opponent. Whether on the open steppes of the Eastern Front, the mountains of Italy or the close hedgerows of Normandy, Allied soldiers learned to fear and respect the German sniper.

WAR IN THE WEST

The British Army was given time to resupply and retrain its battered forces after the disasters of 1940. The importance of sniping was belatedly realized and training schools were set up and efforts made to procure suitable sniping rifles. The Pattern 14 from the First World War remained the standard sniping rifle during the first half of the war. Officially designated the Rifle No. 3, Mark I(T), the P.14 was fitted with a telescopic sight (the P.18) mounted directly above the receiver.

In 1942 a new rifle was introduced, a sniper version of the .303 Rifle No. 4, Mark I, which was replacing the old SMLE as the infantryman's rifle. Designated as the Rifle No. 4 Mark I(T), these were superior models taken from normal production batches and fitted with telescopic sights, most commonly the No. 32 with 3-power magnification. A wooden cheek-piece was screwed into the stock to give the firer a proper stock weld when looking through the telescopic sights. A reliable if unspectacular weapon, the No. 4 was favoured by most British and Commonwealth snipers.[23]

Along with his rifle the British sniper was equipped with a scout regiment telescope with a 20-power magnification, an invaluable aid for observation at long range. The telescope would be carried by the spotter in the standard two-man sniper team, while the sniper would have a pair of binoculars. A prismatic compass and watch were also issued as aids to cross-country navigation. The normal allotment of ammunition for a sniper going out into the field comprised fifty rounds of .303 ball, preferably from a superior batch of ammunition, along with five rounds of tracer (seldom used) and five rounds of AP ammunition (for use against lightly armoured vehicles, vision slits of tanks and machine-guns). Two grenades were issued for personal defence, although some men were wary of carrying such lethal objects when there was always the possibility of safety pins working loose when crawling over rough terrain.

Other equipment included a Denison smock – a windproof and waterproof over-jacket originally designed for airborne troops – which was ideal for snipers. It came in a camouflage pattern, was roomy with plenty of pockets and was fitted with a tail-like cloth flap that went between the legs and press-studded on to the front of the jacket, preventing it rucking up when the sniper was crawling backwards. One or more face veils were issued, primarily for covering the face and hands and breaking up the general body outline, although they were also useful when erecting small hides. Tubes of camouflage cream in green and brown for disguising hands and face were applied in training, but in the field soot and earth were more common forms of skin disguise. The British sniper rarely adopted the complex personal camouflage schemes popular with the Germans. Ghillie suits were occasionally worn, but a smock, face veils and a few pieces of natural camouflage were more usual.[24]

Although the British sniper was adequately armed and equipped, the quality of instruction was not uniform. In 1940 a training course was set up at Bisley to be followed by others in Scotland and Llanberis in Wales. Captain William Jalland of the Durham Light Infantry attended one of the Llanberis courses, which lasted three weeks and was run by the Lovat Scouts. Jalland and his fellow officers (and sometimes NCOs) were part of an instructor's course; on completion they went back to their parent battalion and trained the unit's snipers themselves. This system – which continues today in the British Army – was an efficient way of swiftly imparting general sniping knowledge to relatively large numbers of men, but with the obvious disadvantage that an officer like Jalland would not have the same level of expertise to pass on to his trainees as he had himself received from the Lovat Scouts.

Jalland was taught the basics of sniping: how to fire a sniper rifle with telescopic sight, the best way to use binoculars and the scout telescope, stalking, fieldcraft and camouflage. Like most men before him, Jalland was impressed by the skills of his tutors: 'The way in which the Lovat Scouts were able to cover an open hill side, and yet get in firing range of us spectators without being spotted by anybody, was absolutely incredible – even though we were using field glasses.'[25]

A more idiosyncratic training was given to Sergeant William Spearman of No. 4 Commando. He attended a marksman's course at Bisley before being despatched to a large estate in Scotland. There he was given some rudimentary training in fieldcraft by a local ghillie before being sent out in the Highland glens to live rough for two weeks with instructions to cull the deer herds. He was told to get as close as possible to the target, relying as much on his ability in stalking as in shooting. After bringing down twenty-two stags, Spearman was pronounced a sniper and sent back to his unit as the NCO responsible for sniping.[26]

Outside Britain, training centres were set up in Palestine, Lebanon, Italy and, once the Second Front got underway, north-west Europe. A basic criticism of British training lay in the fact that it was based around the conditions experienced in the First

World War which, although sound for imparting the basic principles of the subject, failed to take into account the greater mobility of action that was characteristic of the Second World War. As a general rule, the most effective centres were those nearest the front and most in touch with the fighting conditions snipers would be likely to encounter. This also allowed instructors to assess their pupils' ability as snipers by taking them on 'practicals' against a real enemy.

A problem for instructors lay in the men themselves. Whereas in 1914 the Army laid great stress on a high quality of overall rifle shooting, this was no longer the case during the Second World War, when the rifleman's place in the tactical scheme of things had been diminished by the advent of highly potent support weapons, notably the mortar and the light machine-gun (LMG). Captain Shore described the ordinary British infantryman's lack of interest in firearms: 'In the main I found the average man's handling of weapons atrocious; he was awkward and bungling, and it was terribly difficult to get into men's minds a true appreciation of the rifle.' In an attempt to improve the situation Shore printed copies of the US Marines' 'Rifle Creed' and distributed one to every rifleman in his unit: 'But', he said, 'it was apparent that only a few men took it to heart.'[27]

As a means of demonstrating the value of aimed rifle fire to a batch of sniper students, one instructor set up a trial between an LMG team and a marksman armed with a sniper rifle: both were to fire at twelve steel plates (six each) which would fall over when hit. In just over thirty seconds, the sniper had his six plates down before the LMG had hit one.[28] The result was testimony to the advantage of accuracy over volume fire.

The fast-flowing war in the deserts of North Africa did not lend itself to sniper actions but once the fighting slowed in the mountains of Tunisia – and in the subsequent campaigns in Sicily and Italy – the need for snipers grew. The exposed mountainous terrain placed a premium on long-range shooting (often well over four hundred yards) which ran against general British practice of attempting to close with the enemy. A sniping officer in Italy gave an account of long-range shooting and a somewhat basic if effective solution to the problem of excess range: 'We found an unsuspecting Boche about six hundred yards away from us, and we could not get any closer to him. So we lined up three snipers together and got them to fire simultaneously hoping that one of the bullets would hit. The hope was fulfilled!'[29]

Intelligent and patient observation was the key to gaining results, especially if the enemy was unwary and did not suspect the presence of snipers. An example of this approach was described by Shore during the Italian campaign. British troops had advanced to take up a position along the River Senio:

> The forward platoon of the unit was in and around a cluster of smallish houses about two hundred yards from the bank of the river. From the roof of one of these houses there was a good clear view of the top of the bank held by the Huns. Snipers watching the bank observed that the Germans changed their sentries every hour with monotonous regularity. At first the Hun was cautious and our snipers withstood the temptation to shoot, hoping that targets would become more favourable when the Jerries had lost some of their caution. Later in the day the hoped-for happened, and at 1200 hours six of the enemy could be seen from the waist upwards. There were four of our snipers on duty and, having their set plan of execution ready, they each selected a Hun and fired. Three of the four Huns fell, and shortly afterwards their bodies were carefully dragged from the top of the bank by their comrades concealed below. At the 1400 hours relief the sentries were again very cautious and the snipers did not get in a single shot. But at 1600 hours two more Huns were sent to their particular Valhalla. It was a long and tedious day for the snipers with only two 'volleys'. But it was a good day's work, five Germans having been accounted for without loss to our men.[30]

The campaign in Normandy was particularly advantageous for the sniper. The succession of embanked hedgerows that formed the *bocage* country of the interior provided perfect cover and concealment. Possessing the advantage of fighting on the defensive, German snipers were highly effective. Fear of snipers came to assume dramatic proportions for British and American soldiers, so much so that almost any rifle fire was put down to their presence. In a survey of his own men, conducted to find out what kind of fire produced the greatest mental strain, Captain Shore discovered that sniping or aimed rifle fire was the most disconcerting, ahead of mortars, shelling and even machine-gun fire.[31] The ability of German snipers – relatively few in number – to cause confusion and dismay can be gauged from this eye-witness account:

> . . . at night [German] snipers crept through the positions, to open fire . . . on parties coming up from the rear. Dozens of bloody little battles were fought behind the forward positions. The snipers were everywhere. Officers, their chosen prey, learned to conceal all distinguishing marks, to carry rifles like their own men instead of their accustomed pistols, not to carry maps or field glasses, to wear pips on their sleeves instead of conspicuously on their shoulders,[32]

The instance of officers (and NCOs) disguising their badges of rank became a vexed question for Allied commanders. When Shore landed in Normandy he instructed his NCOs to cover their rank chevrons with pieces of cloth, while he and fellow junior officers wore roll-neck jerseys to hide the distinctive collar-and-tie. 'A few hours later,' Shore wrote, 'I was told by my colonel that he had ordered my NCOs to uncover their chevrons immediately, and that his officers must wear, and show, collars and ties. If we were to die, he said, we must die as officers!' While one can have little sympathy for the Blimpish attitude shown by Shore's CO, it was important for officers to be immediately recognizable to their men, especially as the constant intake of new drafts made personal recognition extremely difficult.[33] Whatever the merits of either argument, it was clear that the sniper was detrimentally affecting the performance of enemy units facing him.

As the fighting in Normandy slowed into positional warfare, British snipers were able to find good German targets. Captain William Jalland of 8th Battalion, Durham Light Infantry, found that his snipers regarded Normandy as ideal country, chalking up good scores against careless German units: 'They came back one day and they'd just shot somebody on the lavatory, which gave them a certain amount of amusement; they'd also caught some Germans queuing up for food, and managed to kill one or two of them.'[34]

A Scottish battalion holding a position near Caen, known as the 'Triangle', were troubled by mortar fire so accurate that it could only have been directed by an observer hiding in woods some three hundred yards opposite. The battalion sniping section had taken up a post in No Man's Land – a wrecked German armoured car – and began the process of locating the enemy observer. The snipers worked on a shift system, one pair of snipers replacing the other pair at noon:

> The midday relief had just been effected and the two snipers going off-duty were crawling back to our lines when a low whistle and whispered injunction halted them and caused them to recrawl the ten yards or so they had covered back to the post. One of the relieving snipers, bringing fresh eyes to the job, had spotted the German observer; he was high in a tree, excellently camouflaged, but he had made the most elementary mistake of lighting a cigarette – no doubt it was a case of 'familiar contempt'. A whispered conversation between the four snipers followed as to the range of the German observer; they could not agree. Two of them said they estimated the range to be between 250 and 300 yards; the others said it was nearer 350 yards. The matter was settled in a rather novel manner. Three of the snipers set their sights at 250, 300 and 350, while the fourth sniper took the binoculars and kept them riveted on the prospective target. When quite satisfied he coolly gave the three men a fire order; the three rifles 'spoke as one' and the Boche came somersaulting to the ground.[35]

The Commando NCO, Sergeant William Spearman, was deployed on the extreme left flank of the Allied position, where he and his fellow commandos saw continuous action for eighty-three days from D-Day onwards. Responsible for reconnoitring enemy positions as well as sniping, Spearman found he enjoyed the thrill and independence of operating solo. When on reconnaissance missions he adopted the somewhat alarming tactic of deliberately exposing himself to view in order to draw German fire from a distance and thereby pinpoint their positions. An expert stalker, Spearman would creep along the hedgerows to shoot the enemy at short range, although this tactic did have its own psychological problems, the fearful Hun aggressor transformed into a simple human being: 'You'd see the German doing recognizable things the other side of the hedge, such as combing his hair in a mirror attached to a tree. These were the sort of people I sniped at and shot. You knew if you'd hit your target because they dropped in a special way. I didn't like doing it a lot of the time.'[36]

The US Army was less prepared for sniping than its British ally. Despite suffering from enemy sniping in Europe and the Pacific, the American high command never committed itself to systematic sniper training, which was mostly left to individual infantry units to find and instruct the troops as they thought best. While the higher shooting standard prevalent among the American civilian population was a useful preliminary to sniper training, the absence of a codified instructional scheme led to a haphazard mastery of the many facets of the discipline.

The US Army still used large quantities of the trusty .30 Springfield rifle, snipers being equipped with the uprated M1903A4 variant. The Springfield was generally popular with snipers, although its Redfield mount and Weaver sight (slightly less than 3-power) lacked the necessary robustness to endure the knocks of battlefield conditions. And as the M1903A4 was not fitted with conventional iron sights, any damage to the scope rendered the rifle useless.

The bulk of the US Army was armed with the semi-automatic .30 Garand M1 rifle. Two sniper versions were manufactured, the M1C and M1D, the main differences between them being the way in which the sights were attached. As the telescopic sights were offset mounted, leather cheek pads were laced on to the stock to allow the sniper proper eye relief and a good stock weld. Funnel-shaped flash cones were fitted on to some of the Garands.

Whether the M1 was a better sniper rifle than the Springfield remained a matter of debate, a slight balance of opinion favouring the bolt-action Springfield.[37]

One of the first and most successful US Army sniper training programmes was conducted by the 41st Armoured Infantry Regiment, under the command of Colonel Sidney Hinds. A five-week sniper course was introduced after the regiment's arrival in Tunisia, producing several batches of trained snipers.[38] In other units, however, sniping did not prosper if the commander showed little interest, and even when the interest was there he would often lack the qualified instructors he needed to turn out properly trained men.

The chief weakness in US Army training methods was the disparity between formal lessons in marksmanship and field conditions. Great emphasis was placed on rifle shooting at the expense of fieldcraft. As the war progressed some ad hoc sniper schools were set up behind the lines, but these tended to conduct short courses in using telescopic-sighted rifles.

The role and tactics of American snipers were much the same as in other armies. They operated in pairs, their main responsibility to dominate No Man's Land and enemy positions directly opposite. Snipers were expected to pick off high-value targets, incuding officers, NCOs and the crews of enemy weapons posts, as well as dealing with enemy snipers. The ability to hit a man in the body at four hundred yards and in the head at two hundred was the minimum standard of marksmanship.

Following the hard-fought yet successful landings at Salerno in September 1943, US troops began the long haul up the length of the Italian Peninsula. Sergeant John Fulcher, a sniper in the US 36th Division, soon found he had an opportunity to use the hunting skills he had learned as a farm boy in Texas. Fulcher and his spotter partner had slipped away from their lines before daybreak and taken up a concealed position behind enemy lines in the hope of finding a suitable target. Fulcher provided a vivid description of his first successful sniping operation:

I was hiding among a jumble of boulders high on the side of a ridge that offered me an escape route out the back. The sun came up and bathed the broad valley below in soft morning light. A dirt road wound through the short brown grass and among a scattering of leafless trees.

I spotted troops coming at the end of the road where it hazed into the horizon. I nudged my partner and nodded in their direction. We continued to squint into the sun as the troops became individual soldiers marching in company formation, like they were in a Berlin parade for the Führer or something. Through binoculars I could tell they were green replacements. Their uniforms were still a crisp grey-green; their jackboots, kicking up little spurts of dust, still shone. They left a cloud of dust hanging in their wake. Apparently, they were on their way to the front.

I looked at my partner. He had his rifle scope trained on them. He looked back at me. He shook his head.

A *whole* company?

Peering through my rifle scope, I could see the oily glisten of sweat on the commander's brow. Most of the time you could tell the officers. They wore the sidearms and were always shouting and waving their arms.

I never much liked officers anyhow.

I nodded at my partner. Let's take them. They're green. Even if they organized an assault, we could be gone off the ridge before they got halfway across the field to us.

My hands remained as steady as when I shot my second deer. As cool as could be, I cross-haired the officer and shot him through the belly. He looked momentarily surprised. He plopped down on his butt in the middle of the road. The report of the shot reached him as he fell over on to his back. He was dead by the time I brought my rifle down out of the recoil and picked him up again in my scope. His legs were drumming the road, but he was dead. His body just didn't know it yet.

The other Krauts were so green they didn't know enough to scatter for cover until my partner got in his licks by knocking down one more. Even then they behaved more like quail than combat troops. They hid in the drainage ditches and in some shell craters, their heads bobbing up. Just like quail. I figured I could have drilled two or three more, but I held my fire. It wouldn't do to be pinpointed, even by green troops.[39]

Unable to locate the snipers, the Germans reorganized themselves and carried on with the march, only to be hit by another American sniper team further along the road. By the time the German company reached the front the troops were thoroughly demoralized, their fighting efficiency dramatically reduced.

In Normandy, US forces found themselves under repeated sniper attack. Like the British they were unprepared for the close fighting characteristic of the *bocage*, where a German sniper or machine-gun nest seemed to dominate every small patchwork field. For the commanders, keeping the troops moving under fire was a constant problem; in many instances a lone German sniper could hold up a platoon attack for hours if heavy support weapons were not available.

The effectiveness of such a sniper was described by a platoon leader in the US 9th Division, whose inexperienced troops suddenly came under enemy fire: 'One of the fatal mistakes made by infantry replacements is to hit the ground and freeze when fired upon. Once I ordered a squad to advance from one hedgerow to another. During the movement one man was shot by a sniper firing one round. The entire squad hit the ground and they were picked off, one by one, by the same sniper.'[40]

As the American troops grimly fought their way through the *bocage* they came to loathe the menace that scythed down their comrades from out of nowhere. An example of this exasperation, which was felt from top to bottom of the command structure, was recorded in a diary entry by the ADC to General Omar Bradley: 'Brad says he will not take action against anyone that decides to treat snipers a little more roughly than they are being treated at present. A sniper cannot sit around and shoot and then [expect] capture when you close in on him. That's not the way to play the game.'[41] As in other wars the sniper could expect little mercy when found with a smoking gun.

The only way to deal with a sniper was to destroy his position by superior firepower from mortars, artillery or tank guns, or send in another sniper. Whatever the chosen method, the Allies never properly eliminated the German sniper threat. Only the eventual collapse of the German line in July 1944, transforming the fighting from static defence to a war of fast-moving armoured manoeuvre, reduced the sniper's deadly influence.

SNIPING IN THE PACIFIC

In Asia and the Pacific the Allies were faced by snipers of the Japanese armed forces; their mastery of camouflage and determination to fight literally to the death made them a redoubtable foe. The Japanese 6.5mm Type 38 rifle was adopted for service use in 1905 and became the standard infantry arm. A sniper variant was developed – the Type 97 – which was fitted with a telescopic sight (2.5-power) and a revised bolt handle. Although considered underpowered by European standards, the Type 97 had a great sniping advantage in its reduced cartridge and long 31-inch barrel: the entire charge burned fully in the barrel, eliminating almost all muzzle flash and smoke. Allied soldiers consequently found it difficult to pinpoint Japanese snipers, especially in the jungle where location of sound was often deceptive. During the course of the war the Japanese fielded the Type 99, a rifle in the more conventional calibre of 7.7mm (.303in). Firing standard ammunition, the sniper version of the Type 99 was provided with a 4-power scope and a folding monopod attached to the front band, intended to provide greater stability.[42]

As much of the fighting against the Allies took place in jungle or other close terrain, accuracy at long range was of less importance to the Japanese sniper than good marksmanship at distances normally under three hundred yards, and sometimes as close as a hundred yards. The Japanese Army had always excelled in camouflage techniques and great attention was paid to personal camouflage. Helmet and body nets were issued to each sniper as a frame to build elaborate schemes of concealment, although in practice snipers in the field adopted a simpler approach when they discovered that the camouflage became too easily entangled in the jungle foliage.[43] Fieldcraft and observation were stressed in training, and the Japanese sniper was adept at moving unseen through most types of terrain. His natural hardiness and patience allowed him to take up positions, seemingly regardless of discomfort, for periods well beyond the normal endurance of a Western soldier.

Japanese sniper tactics were broadly similar to European practice. The most important task undertaken by the sniper was to kill or capture high-value targets, notably officers or enemy snipers. Thereafter he was instructed to neutralize or destroy enemy installations which might affect his own unit's mission; to attack enemy support weapons and the crews serving them; and, lastly, to take on any targets of opportunity that might come within range.[44]

A distinctive feature of Japanese sniping was the use of trees as firing platforms. In some cases small tree chairs would be hauled up into the higher branches; in others the sniper would be tied into position, which prevented him falling out of the tree if shot, thus informing the counter-sniper team that they had scored a hit. As an aid to clambering up and down the trees, the sniper was issued with climbing spikes. To Allied and German snipers the use of trees was discouraged (although observers used them regularly, and were often mistaken for snipers) because they became a death trap if the sniper was discovered.

Such scruples meant little to the Japanese, and snipers regularly let themselves be overrun by the enemy in order to loose off a few well-placed shots before being caught and eliminated. Although such fanatical bravery had much to commend it, it was less useful for a sniper, where discretion is as important as courage. The British sniper officer, William Jalland, rightly listed 'a strong element of self-preservation' among the necessary requirements for a good sniper.[45]

None the less, throughout the Pacific island campaigns, Japanese snipers were able to pin down large numbers of men and create confusion and fear out of all proportion to their numbers. An example of this was provided by the action of

Japanese snipers on the Kwajalein atoll in the Marshall Islands chain, which the US 7th Infantry Division was trying to clear in January–February of 1944:

> On the last day of the five-day battle for the atoll, Company F, 32nd Infantry Regiment, found themselves pinned down by sniper fire. The men could not tell where it was coming from. The bullets paralyzed the men. Then they tried to dig deeper into the sand or cover themselves with palm fronds in an attempt to hide. For an hour the company 'clung to the earth' just 150 yards short of the end of the atoll. One by one ten soldiers were hit by the sniper fire and each time the medics risked their lives crawling forward to the wounded and dragging them back. The 'will to go forward' had vanished. Only the arrival of tank support as a shield from the sniper fire energized the men from Company F to get up and move out towards their objective. They raked the ground in front of them with constant BAR [Browning Automatic Rifle, the squad light machine-gun] fire, flushed out the snipers, knocked out enemy bunkers and cleared out the last opposition on the atoll.[46]

Away from the exposed coral atolls of the Pacific, Japanese snipers could still be effective, as they demonstrated in the dense jungles of New Guinea. The 1st Battalion of the 163rd Infantry Regiment, US 41st Division, was particularly troubled by snipers while fiercely contesting ground against the Japanese in the Musket perimeter during December–January 1942–43. The jungle there was particularly thick, and the stealthy Japanese snipers would take up position in the trees surrounding the perimeter. The divisional historian explained the situation facing the hard-pressed Americans:

> From a tree almost anywhere around our oval perimeter, a Jap sharpshooter could choose a Yank target who had to leave his water-soaked hole. The range could be all of 200–400 yards. The keen-eyed sniper could steady his precision killing tool on a branch and tighten the butt to his shoulder. He could take a clear sight picture and squeeze the trigger. All 1/Bn might hear is a Jap .25-calibre [6.5mm] cartridge crack, like a Fourth of July cap sparked on a stone. Then a Yank cowering in a hole might

hear the prolonged dying groan of a man in his next squad. Or long after a deadly silence, he might find his buddy a pale corpse with a deceptively small hole in his forehead.[47]

As the Japanese snipers grew in numbers the troops in Musket perimeter began the long and arduous task of eliminating them. The Americans were nothing if not thorough, deploying a mutually supporting set of counter-measures. Two-man counter-sniper teams manned the forward defences, while other teams clambered up to the treetops and moved Tarzan-like through the jungle, using ropes and collapsible ladders. They were able to hold off the immediate Japanese threat, while other teams of two or three soldiers would set off into the jungle on offensive operations, guided towards their targets by the soldiers in the trees.

Through coordination of the various counter-sniper elements, the Japanese positions were located and their snipers killed by American small-arms fire. While the battle to subdue the snipers was under way, the 1st Battalion secured three 37mm anti-tank guns. Putting aside their solid-shot ammunition the gunners loaded up with 'grapeshot' and blasted the trees wherever a Japanese sniper might be stationed. Even if the tactic wasted ammunition, it was a boost to morale. These measures radically cut down the effectiveness of the Japanese sniping effort, prior to the main US assault on 16 January 1943 which drove back the Japanese from the Musket perimeter.[48]

British and Commonwealth soldiers fought similar battles with snipers in the jungles of New Guinea and Burma. Once the troops realized that the Japanese sniper was no superman and could be silenced through the systematic application of counter-sniper tactics, the battle was largely won. During the course of the war, as the quality of Japanese infantry declined, the British began to achieve notable sniping successes of their own. An official report stated that the combined forces of two brigades of snipers (forty-eight men) were responsible for killing 296 Japanese over a period of two weeks; British casualties comprised two men killed and one man wounded in the finger.[49]

Australians made good snipers, the best taken from the ranks of kangaroo hunters. To be a successful 'roo hunter a man had to be an excellent shot, for unless the bullet hit home precisely the valuable pelt was spoiled; also, a wounded, thrashing kangaroo alerted the rest of the herd while a clean kill left them undisturbed. Australian marksmen had much experience of using heavy barrelled .303s, which stood them in good stead when stalking Japanese targets.

One hunter-turned-sniper chalked up a score of forty-seven Japanese on the island of Timor, but only claimed twenty-five on the basis that 'in my game you can't count a 'roo unless you see him drop and know exactly where to skin him'.[50] Operating in independent companies on Timor the Australians fought a guerrilla campaign against the Japanese, inflicting a reputed 1500 casualties for the loss of forty men. As a result, large numbers of Japanese reinforcements had to be diverted to Timor.[51]

The US Marine Corps was generally better disposed towards the concept of sniping than its US Army counterpart. This was largely a result of the Corps's determination to develop basic infantry skills to a high level, sniping considered as one of these skills. Yet this commitment was far from consistent and varied according to season, waxing in wartime but waning dramatically once peace was declared.

Within the Marine Corps, George O. Van Orden was a leading proponent of increasing sniper capabilities, and in 1942, along with Calvin A. Lloyd, he wrote a report advocating the adoption of a complete sniping philosophy for the Marine Corps.[52] A key point of Van Orden's report was the recommendation that the Marines should use a rifle designed for sniping, or at least one with a hunting pedigree, rather then merely modifying a standard service rifle. In retrospect, this was an obvious suggestion, but at the time it was viewed with grave suspicion by most ordnance officers.

After considering the weapons available in the United States, Van Orden put forward a claim for the Winchester G7044C, an accurate heavy barrelled bolt-action hunting rifle, which would be used with a Unertl 8-power scope and match-grade ammu-

107

nition. The Marines already had some experience of the Winchester M70 (at the request of Colonel Meritt Edson of Marine Raiders fame), but in the end they played safe and rejected the Winchester in favour of the Springfield and Garand models, on the basis that they did not want to inject another rifle into the supply chain.

Marine snipers armed with the Springfield were given the M1903A1 model, fitted with the 8-power Unertl scope which, despite being somewhat fragile, did provide good magnification for long-range shooting. In keeping with the haphazard training schemes set up by the wartime Allies, sniper instruction in the Marine Corps largely depended on circumstance. In some instances, a Marine who was considered a good shot was given a telescopic rifle and the title 'sniper' on his arrival in the field. Elsewhere, there was more formal training, with sniper schools set up in Camp Lejeune, North Carolina (December 1942), and at Green's Farm, San Diego, California (January 1943).

The school at Green's Farm came under the command of Lieutenant Claude N. Harris, a noted marksman and winner of the 1935 National Rifle Championship. Harris ran a series of five-week courses, where fifteen-strong teams were instructed in a broad-based syllabus which included marksmanship, camouflage, observation, fieldcraft, map reading, military sketching and the interpretation of aerial photographs. This reflected the Marine belief that the sniper should be capable of fulfilling a reconnaissance role, hence their official title of 'scout snipers'. After graduation they were allotted to field units, three to each company; two would form the usual sniper-spotter team while the third man would act as a reserve in case of casualties or illness.[53]

Marine snipers were mainly deployed in a counter-sniper role, as there were relatively few opportunities for sniping directly at the enemy. In extreme instances, such as during the battle for Tarawa (November 1943), the scout-sniper platoons were used as shock troops, landing with the assault engineers ahead of the main attack, their function to eliminate Japanese waterfront defences. Only when the battle moved 'inland' did the scout snipers act in a more conventional manner, defending the regimental command post against the threat of Japanese snipers.[54]

In subsequent island battles snipers led the attack on Japanese strong points, especially those which were inaccessible to artillery and air bombardment. During the invasion of Saipan in July 1944, the 2nd Marine Division used snipers armed with Springfields and reported excellent results. In the sustained battle to secure Okinawa (August–October 1945) snipers were called upon to root out the remnants of the Japanese defending force, fighting on in the wooded hills of the island. In one of many engagements on Okinawa, Private David Webster Cass Jnr knocked out a machine-gun emplacement, holding up a Marine advance, at a range reputed to be at least 1200 yards.[55]

Despite the exploits of snipers in these operations, the Marine Corps Headquarters remained unconvinced of the overall effectiveness of sniping. The focus of military attention lay upon the more obvious and larger-scale successes of the Pacific War, such as amphibious landings and air support. On a cynical level, an interest in sniping did little to further an ambitious officer's career. Even before the end of the war, on 23 April 1945, the Operations and Plans Section had dismissed sniping from a role in post-war Marine training, and even went as far as authorizing the Quartermaster to dispose of excess Unertl-scoped Springfield rifles.[56]

Marine Corps doubts were echoed by the US Army in the Pacific theatre. In a report prepared for US Army Ground Forces, Pacific Ocean Areas, dated 5 January 1945, the HQ of I Corps noted that 'Sniper training has not been prescribed for I Corps for over a year, as their employment has not been considered practical.' An interview with the Headquarters of the US Eighth Army brought a similar response: sniper training was found to be unnecessary and that marksmen at company level would be assigned telescopic rifles when, and if, the need arose. In other words there was no requirement for a tactical doctrine for snipers, nor an Army-wide training programme. The report concluded: 'Specific training of snipers is not at present being carried on by units assigned to this theater. The individual selection for sniper missions of expert riflemen within small units has been found sufficient and is the method currently employed.'[57]

This attitude became standard throughout the US military and sniping went into a steady decline after the Second World War. In Western Europe, a similar lack of interest in sniping prevailed. As a defeated nation, Germany had limited provision for sniping, and when the Bundeswehr became an effective force in the 1950s it was understandably sidelined by an obsession with armoured vehicles as part of the defence of West Germany against the tanks of the Warsaw Pact. The Soviet Union, by contrast, maintained its snipers as a standard element of every Red Army company. In Britain, the Army dropped sniping in the immediate post-war years, only to readopt it in fits and starts as the wars of colonial departure drew attention to a sniping requirement. Only the Royal Marines, who had exclusively taken over the commando role, continued and developed sniping. Subsequently they would become the standard bearers for other Western armed forces.

PART THREE

THE MODERN SNIPER

'People . . . think you go out and hide and blow everybody asunder. We don't do that. We pick our targets – leaders, operators of crucial weapons, communications people. We don't go out and indiscriminately kill.'

Carlos Hathcock (*Chicago Tribune*, 7 Sept. 1986)

CHAPTER SEVEN

THE KOREAN WAR

When communist North Korean troops suddenly invaded South Korea on 25 June 1950, the United Nations was swift to react, assembling a multinational force to come to South Korea's aid. Under American leadership (the US armed services supplied the vast bulk of the force) the perimeter around Pusan was stabilized, before an amphibious landing at Inchon sent the North Koreans reeling back in disorder. While the UN troops fought their way up the Korean peninsula, the war was transformed by the arrival of China on the side of North Korea in October 1951. The outnumbered UN forces were pushed back, until they were able to stabilize the line near the 38th Parallel, where the war degenerated into an affair of localized attacks and counter-attacks. Talks between the sides began, although it was not until 27 July 1953 that hostilities ceased with the signing of an armistice at Panmunjom.

The barren mountains and harsh extremes of climate ensured that the war was especially onerous for the fighting soldier. The bitterly cold winters made sniping difficult, and at times virtually impossible; lying in wait in exposed positions led to hypothermia, and just holding a rifle without gloves caused cold burns. Yet the Korean War proved the need for sniping. The open terrain and the trench-bound nature of the latter part of the war enabled snipers to operate effectively. In some respects the conflict took on the characteristics of the First World War, and the battle for domination of No Man's Land became a challenge for UN commanders faced by an enemy well equipped with snipers.

The Americans were poorly prepared for the sniper battle. The Army and even the Marine Corps regarded sniping as a peripheral activity. Experience in the Second World War had left

113

them unconvinced of the sniper's potential, and the demands of warfare in the atomic age had turned attention away from such a basic infantry skill. Although a telescopic-sighted rifle was authorized for issue at squad level, no attempt was made to develop a sniping doctrine. Sniping was left to the unit commander's discretion, and in most instances it was allowed to languish. Thus, when American troops found themselves under sniper attack they were forced to improvise. Sergeant Bill Krilling's experience was not untypical: 'A single incident sparked my interest in sniping. One day after a close friend was killed by a communist sniper, while moving about the area, the man sent to recover my dead friend was also killed by the same sniper. I decided there and then to become a sniper.'[1]

The American 'can-do' approach could be highly effective, as was demonstrated during house-clearing operations by US Marines in Seoul. A Marine armed with a Browning Automatic Rifle (BAR) worked alongside a sniper with a Springfield M1903A1. The BAR man would let off a magazine at a suspected enemy position; the North Korean soldier would then look over the wall, ready to reply to the incoming fire, only to be hit by the well-sited Springfield-armed Marine sniper. But improvisation was ultimately a limited option, and often the equipment, and the men using it, were not up to the demands of sniping. During the war, a report drawn up from the activities of the 9th Infantry Regiment, 2nd US Infantry Division, made this damning conclusion: 'Many failures with present equipment occur due to lack of proper ordnance support and the individual handling it not being trained in its care and use. Present sniper equipment is not being utilized in all cases as intended, but is frequently issued to a man that would like to "carry it".'[2]

The equipment issued to snipers was of Second World War vintage: M1C/D or M1903A1/4 telescopic rifles. Already past their prime, years in storage had decreased their effectiveness, and snipers complained of poor trigger pulls, badly designed stocks, and indifferent quality control over the M1903A4s. Their main criticism, however, was reserved for the M81 and M82 sights, whose limited magnification, only 2.5-power, was inadequate for the long ranges encountered in Korea. A further complaint was the failure of the rifles to hold their zero, so that they had to be constantly reset on shooting ranges back from the front line.[3] The Marine Corps's Unertl 8-power scopes also came in for criticism; although magnification was good, its very limited field of view made target acquisition difficult, and the design was not considered sufficiently soldier-proof.

During the war, the Marine Corps Equipment Board issued a test report on sniper equipment.[4] The report recommended that the Marines should adopt the M1C rifle, declare the 1903A1s obsolete and discard the Unertl sight and mount in favour of a Griffin and Howe mount and a 4-power Stith (Bear Cub) telescopic sight. Renewed attempts to have the Winchester M70 introduced as a sniper rifle were firmly quashed; to operate efficiently it would require match-grade ammunition, but to include another ammunition type was considered undesirable. Once again the quartermaster won the battle over the sniper.

The report accepted that ammunition quality was not high and recommended an overall improvement in cartridge manufacture for all M1 rifles. The standard .30 calibre ball ammunition used by snipers was acceptable for short-range shooting but lost its accuracy at ranges in excess of five to six hundred yards. While this had not been a major problem in the Normandy *bocage* or the Pacific-island jungles, in Korea this lack of long-range accuracy was regularly exposed.

Some idea of how American front-line troops dealt with the Chinese and North Korean snipers facing them was provided in an account of the experiences of a commanding officer of the 3rd Battalion, 1st Marines.[5] Having just assumed command, the officer went to inspect his positions. Moving to a forward bunker he looked through the embrasure with his binoculars to scan No Man's Land. Seconds later a sniper's bullet smashed into the binoculars, throwing them to the ground and leaving him with a gashed hand. Shaken by the incident the colonel 'reflected that it was a helluva situation when the CO could not even take a look at the ground he was defending without getting shot at. Right there and then he decided that something had to be done about that enemy sniper.' His solution was to select and train a body of crack shots to take on and defeat the enemy.

The colonel first ascertained that there was an adequate number of telescopic rifles held in the supply section. More to the point, he located a gunnery sergeant who was both an experienced combat soldier and a crack shot with experience on Marine rifle-shooting teams. The 'gunny' then took over the selection and training programme, outlining his requirements to each of the company commanders: good infantrymen with the quality of patience – essential for a man who had to spend hours holding a position while at the same time maintaining full alertness. After an initial interview with the sniper candidates, the gunny selected six two-man teams from each company.

A training range was assigned in the battalion rear area and the students were put through a stringent three-week training course. The men used either M1s or Springfields, according to preference, as well as receiving instruction on the .50-calibre machine-gun, firing on single shot with a telescopic sight attached. The course concentrated on marksmanship and was tailored to the immediate tactical situation, to take on the enemy in the lines opposite them. Comprehensive sniper training was a luxury that would have to wait for the men of the 3/1 Marines.

After graduation the snipers went back to their companies to take up camouflaged positions in the front line. There they found the communist snipers had the upper hand. The Marine snipers set about reversing the situation. As a means of encouragement, a case of cold beer was to be awarded to the first men of each outpost who gained twelve kills in a week. The snipers received enthusiastic support from their gravel-crunching comrades, who became expert in locating communist sniper positions.

The turn-around in Marine fortunes was swift and dramatic. The CO recalled that, 'In nothing flat there was no more sniping on our positions. Nothing moved out there, but that we hit it.' Having heard of the deployment of the sniper unit, the divisional commander came up to inspect progress. To his amazement he was able to walk the length of a company frontage armed only with his walking stick, safe from enemy fire. A gratified major-general exclaimed to the company commander, 'What we need are more snipers on this front!'

These experiences were amplified by those of the 2nd Battalion, 5th Marines, holding a section of the line in April 1951.[6] The officers of the battalion found themselves constantly frustrated when trying to hit good targets at medium and long rifle range; collectively, they agreed that they must raise a sniper platoon. The platoon was placed under the supervision of the executive officer, a former commander of a reconnaissance company during the Second World War. Reflecting his experiences, the platoon was raised to perform both a scouting and sniping function. The platoon was made up of three squads, each of which contained four two-man sniper teams.

Training time was shared between the rifle range – improvised behind the battalion command post – and schooling in fieldcraft and scouting, as well as more formal classroom lectures. When the 'Exec' considered his students were ready they were formed into a sniper platoon under the command of Lieutenant Gil Holmes, a young replacement officer and Marine Corps team shot.

The battle front facing the 5th Marines was more open than that of the 1st Marines, thereby allowing the snipers to operate well forward of their own MLR (Main Line of Resistance). But, wary of a sniper team being caught by superior enemy forces, each team was given fire support in the form of an infantry rifle squad. This was a departure from classic practice in which snipers operated on their own, but in the circumstances was a useful measure – and one that would be repeated during the Vietnam conflict.

On 21 May 1953 the Marine snipers had their first chance to demonstrate their skill in a large-scale action. The 3rd Battalion had taken up a blocking position to counter a Chinese assault; the Americans could see the enemy forces, about a thousand yards away, advancing across their front to attack a battalion on their left. In order to discover the exact Chinese positions a patrol was despatched from D Company's front, comprising two rifle squads and four sniper teams under the overall command of Lieutenant Holmes. The snipers were to adopt a scouting role with the rifle squads acting as escorts.

Making full use of the undulating terrain the Marine patrol advanced cautiously towards the Chinese troops. After covering three hundred yards Holmes decided to leave the two escort squads to hold a good covering position on a rise, leaving the snipers to crawl forward on their own, in the hope that they would not be spotted. Despite the care the US Marines took in closing with the enemy, Lieutenant Holmes recalled afterwards how surprised he was that he and his men were not detected. Holmes wrote:

> We finally reached a ridgeline roughly parallel to the entrenchments on [Hill] 719 from which we could look across and see a couple of Chinese wandering around the area in an unconcerned manner. I would say that the range from our position was about four hundred yards. We were all spread along the ridgeline in a loose skirmish line with sights adjusted, waiting for remunerative targets. I finally spotted three at the same time and gave the word to cut loose. That really did it! I had no idea what a hornet's nest they had over there. They came running out of their bunkers along the trench to their battle stations, but it soon was obvious they were rather fouled up.

> They tried to set up a machine-gun to our direct front and one of my boys knocked off the gunner. When they finally got the gun in action they opened up on an area at least two hundred yards from our left flank. Some joker, evidently the company commander, was running around like a madman trying to square things away, but his people were crumbling all around him under a steady stream of the well-aimed fire of our sharpshooters.

> Soon after we opened fire, D Company called us back in. I stuck my neck out and held the position for another fifteen minutes after receiving the order because we had good shooting and the Chinese just couldn't seem to get squared away. They returned fire, but it was ineffective – they didn't seem to have a fix on our positions. Approximately three-quarters of an hour after we broke contact and commenced our withdrawal from

719 we were safely back inside our own lines without spilling a
drop of Marine blood – it was a good day![7]

The inability of the Chinese to locate the source of the Marine
sniping reflects a common problem among inexperienced troops
– especially when panicking under fire – and one which snipers
have been able to exploit to good effect.

The snipers from both Marine battalions were keen to extend
the range of their shooting and, realizing the shortcomings of
their .30-calibre weapons and ammunition, they looked to the far
larger .50-calibre round. The .5in M2 Browning heavy machine-
gun had been developed just after the end of the First World
War, and was produced in both aircraft and infantry versions.
An air-cooled weapon, the infantry M2 was unusual in having
the option for both automatic and single-shot fire, a feature that
brought it to the attention of snipers.

As a machine-gun the M2 was not particularly accurate for
sniping purposes, yet its powerful bullet enabled targets to be
taken on at ranges of up to 1200 and, in certain cases, over 2000
yards. The sniping 'Big Fifties' – fitted with telescopic sights –
had been used in a limited manner during the Second World
War. In Korea, a greater interest was shown in them, even if
their bulk and weight made them unsuitable for sniping except
from fixed positions.

Experienced snipers – including those from the battalions of
the 1st and 5th Marines – wanted a single-shot .50-calibre weapon
specially prepared for a sniping role and with appropriate match-
grade ammunition. For this they would have to wait more than
two decades, but during the Korean war various experimental
single-shot weapons were tested. Most were based around the
old Boys anti-tank rifle, which had been developed in Britain
prior to the outbreak of the Second World War, before being
discarded when tank armour was found to be too thick for
effective penetration. Originally chambered for .55-calibre, the
American snipers converted the Boys to .50-calibre with an M2
barrel, plus a telescopic sight. In keeping with many other lash
ups the results were far from perfect, but reasonably accurate
shooting at ranges of over a thousand yards was possible. A

significant factor affecting performance was the variable quality of the ammunition; to rectify this, some enthusiasts went so far as to build their own loading presses, reloading the cartridges and checking the bullets for consistency.[8]

A pioneer of .50-calibre sniping, Lieutenant-Colonel Frank B. Conway, conducted extensive trials during the late 1940s and 1950s. Instead of the Boys, he used the German PzB39 anti-tank rifle as his base for conversion to .50 calibre. During test firing he found he could achieve favourable results at distances of up to 1400 yards, and on the ranges at Fort Bliss, Texas, an adobe shack was fired on from a range of 2800 yards. Conway recalled that 'we could usually place the second or third round through a small window'.[9] Despite these impressive results, official interest in the Big Fifties did not extend to developing a single-shot rifle for regular sniper use. Interest was renewed during the war in Vietnam, but as in Korea this was only a temporary phenomenon, and it would not be until the 1980s that tailor-made .50-calibre sniping rifles would become a standard sniping weapon.

Towards the end of 1951 the success of Marine Corps scout snipers prompted the 1st Marine Division to codify its practices in a document outlining the tactical deployment of its snipers.[10] The basic unit was the two-man observer-sniper team, equipped with sniper rifle and 7×50 binoculars, both men trained to carry out either role. In support of the snipers was a four-man rifle team, providing security for the scout snipers. Examining the most suitable techniques for deploying snipers, the document noted:

> The scout-sniper patrols have been stationed on prominent noses and ridges forward of the MLR from which they are able to view the enemy. The patrols move quietly along concealed routes of approach to their positions, generally during the hours around sunset and dawn. The teams are most effective in the early morning hours when they remain concealed in the shade from the morning sun shining into the eyes of the enemy.
>
> On occasion they have been able to see and direct fire on the enemy under a blanket of fog which obscured observation from

higher points and gave the enemy a false sense of protection. Snipers have been able to operate with success at ranges up to 1000 yards; however, 600 yards is considered the optimum range. Alternate positions are prepared so that the patrol may shift if its positions are discovered; to date because of careful attention to stealth and camouflage, only one of these patrols has received counter-sniper fire. The best results are obtained by employing the same team in the same position so that they become familiar with the terrain, and are able to detect new targets more readily.

Taking advantage of their advance positions the scout-sniper patrols were a valuable tool in obtaining tactical reconnaissance. So important was this information that the patrols were equipped with land lines to telephone information back to the company command post. Obviously cumbersome – and anathema to the basic sniper tenet of stealth and operational freedom – it appeared to work in Korea. From the company CP, telephone lines were laid to all the supporting arms – artillery, mortars and heavy machine-guns – and so intelligence from the snipers a thousand yards beyond the MLR could be acted upon in a matter of seconds. 'Many times,' the report concluded, 'targets which otherwise would never have been seen, were reported by the farseeing scout snipers and taken under fire immediately by the co-ordinated fire-control system. Often these targets were visible for a short time only, and with a less direct system, would not have been taken under fire.'

Despite the success of sniping in Korea, when the war ended, commitment waned – much as it had done following the Second World War. For the higher military authorities, sniping remained a battlefield expedient, which in peacetime only got in the way of ostensibly more important matters. Yet at lower levels within the armed forces there was a belief in sniping and a body of knowledge that would prove invaluable in the next war.

CHAPTER EIGHT

WOUND BALLISTICS

Studies and histories of sniping understandably concentrate on rifles and ammunition, for without them the sniper cannot operate. Yet these are merely a means to an end: the death or incapacitation of the enemy soldier. What a bullet does to the human body is a complex business, dependent on many variables, some unquantifiable. None the less it is an essential part of any analysis of sniping, and the expert sniper will be aware of the properties of his ammunition and its effect on human targets.

The low-velocity large-calibre ammunition fired by the smooth-bore muskets of the eighteenth century was – when it hit its target – capable of causing severe wounds. The heavy lead musket ball deformed on impact to create a flattened slug approximately one inch in diameter, which retarded swiftly so that the ball usually remain lodged within the tissue. The transfer of energy from bullet to body was correspondingly efficient. The development of the rifle-musket in the early to mid-nineteenth century provided greater accuracy, and because the heavy Minié bullets travelled at increased velocity, the wounds inflicted were more severe.

Surgeons operating in the American Civil War witnessed the damage caused by bullets fired from Minié-type rifles, such as the Enfield, Springfield and Whitworth. In 1877, the eminent surgeon Professor T. Longmore gave this account of their effect on the human body:

> If a modern rifle bullet, armed with its full force, strikes a hard and powerful long bone, like the femur for example, near the middle of its shaft, it is broken into fragments of various shapes and dimensions often too numerous to be counted. A large proportion of these fragments are driven violently in various

122

directions, and thus are converted into secondary missiles. A huge hollow is formed inside the limb, which, when it is fully laid open and the effused blood washed away, offers to view a mass of lacerated muscle intimately mixed with sharp-pointed and jagged-edge splinters of bone. With all this extensive destruction within the limb, the external aspect of the wound through which the bullet first entered may exhibit nothing more to view than a small opening into which the top of the little finger enters with difficulty.[1]

Longmore's findings were part of a general and systematic attempt to understand the nature of gunshot wounds. Detailed studies were conducted in America, Britain, France and Germany, building up a body of knowledge that gave wound ballistics a scientific basis. As it was impossible to gain a full scientific understanding of bullet wounds on live human beings, other methods were adopted. The Prussian War Ministry conducted trials in which bullets were fired at cadavers (preserved in a formalin solution) at ranges varying from twenty-five to two thousand yards. When living tissue was required, anaesthetized animals – mainly pigs and goats – were found useful in simulating live human wounds. More frequently, however, bullets were fired directly into blocks of clay, soap or a gelatin solution – media which demonstrated many of the properties of human tissue. In the 1960s computers were utilized to describe the ballistic characteristics of bullets in almost any given situation. These methods, employed alongside clinical studies of actual bullet wounds, provided sufficient information for the emergence of general principles of wound ballistics.[2]

The essential property of a rifle bullet in flight is its velocity. Primarily, high velocity is intended to ensure accuracy, but greater wounding power is an important secondary factor. Experts vary in their definition of high velocity,[3] but it can be taken that all military rifles fire high-velocity bullets. There are three separate measures of velocity: initial (or muzzle), the velocity of the bullet as it leaves the rifle barrel; impact, the velocity of the bullet as it hits the target; and residual, the velocity of the bullet as it exits the body (assuming that the bullet is not lodged in the tissue).

The difference between residual and impact velocities is a key element in determining the severity of the wound: the greater the retardation of the bullet, the greater the absorption of its energy and the greater the potential injury. A 'weakness' of some full-power, high-velocity ammunition is the bullet's tendency to slice straight through human tissue without releasing the bulk of its energy (although this 'weakness' is only relative; damage to head or torso by high-velocity bullets is almost always massive).

The rifle bullet is an inherently unstable object, unlike, for example, an arrow, which naturally tends towards a stable point-first course. Long and thin, with a pointed nose, the rifle bullet is the optimum shape for cutting through air at over the speed of sound. When in motion it has a tendency to oscillate on its long axis – a phenomenon known as yaw – but this is controlled by the rifling in the barrel which imparts an extraordinary fast rate of spin. This gyroscopic action keeps the bullet on target as it describes a normal ballistic trajectory. Yet the stability provided by spin cannot be maintained in a denser medium than air; when it strikes human tissue (800 to 900 times denser) the bullet immediately becomes unstable, and it tumbles. The degree of bullet instability varies considerably according to its design, and the density and elasticity of the tissue it is travelling through.[4]

For the military sniper the science of wound ballistics is simplified into two requirements: to kill the enemy or to render him inoperable for further military action. But for police, para-military and army counter-terrorist units, far greater precision in target incapacitation is often necessary. In certain situations – particularly where hostages have been taken – there is a precise requirement for immediate incapacitation.[5] This is a literal definition which, for example, would prevent a hostage-taker having the physiological ability to press a trigger (even as a reflex action) the moment the sniper's bullet penetrated his body.

Immediate incapacitation is only achieved by destroying the spinal column above the second or third thoracic vertebra (above the shoulder blades) or inflicting severe injury to the brain stem

124

(medulla oblongata) or the neural motor strips on either side of the brain. When a bullet strikes these areas the target will collapse instantaneously – as if poleaxed. Except at short range – for this purpose a hundred yards or less – immediate incapacitation cannot be guaranteed from even the best marksman. Consequently, this is a major reason why paramilitary snipers must operate at much closer ranges than conventional snipers.

The next level down in the index of injury is rapid incapacitation, whereby the target is either killed or rendered completely incapable within the space of a few seconds. A bullet hitting the heart, for example, causes unconsciousness and then death within ten to fifteen seconds. Similar forms of rapid incapacitation follow from shots to any part of the brain and spinal cord, the liver, major blood vessels and the kidneys. Only a very few, extraordinary men are capable of any purposeful action after being hit full-on with a high-velocity rifle bullet in the head or torso.

When the bullet strikes the target it causes injury in three main ways.[6] The first of these mechanisms of injury is the laceration and crushing of tissue as the bullet cuts through the body, creating the permanent wound track. If the wound track passes through a major organ or severs important blood vessels, then the injury will be appropriately devastating. All bullets which hit the body (whether low or high velocity) will carve out a permanent wound track, although only the tissue in direct contact with the missile will be damaged. The size and depth of the permanent wound track will depend on the tissue encountered, the energy of the bullet and its relative stability.

The second mechanism of injury – the hydraulic shock wave – only occurs with high-velocity missiles travelling above the speed of sound. As the bullet begins to cut its wound track it compresses the tissue in front, sending out a shock wave at a velocity approximately the speed of sound in water (4800fps). Although the shock wave may last no more than a millionth of a second, it can cause damage at distances removed from the wound track, especially when conducted along fluid-filled avenues such as veins and arteries. While denser tissue is better able to withstand the hydraulic shock wave, organs such as the liver, spleen or brain – containing incompressible tissue with a relatively high fluid consistency – are particularly vulnerable.

The third mechanism of injury is temporary cavitation, produced by high-velocity missiles, and is extremely destructive of human tissue. As the bullet enters the body its tremendous level of energy is absorbed by the tissue around it. The energy is so great that the tissue is forced forwards and outwards; this motion gains its own momentum and continues after the bullet has passed through the body, to form a large cavity, up to thirty times the size of the projectile. The cavity subsequently collapses in a series of pulses, to leave the permanent wound track.

'Soft tissue will be pulped,' writes the distinguished military surgeon, Professor M. S. Owen-Smith, 'small blood vessels disrupted and bone may be shattered without being hit directly.'[7] The disruption of nerves may be sufficient to cause paralysis and the pressure on blood vessels at some distance from the wound may bring about thromboses. Such is the destructive nature of cavitation that conventional high-velocity bullet wounds have regularly been confused with the use of explosive projectiles.

The deleterious consequences of cavitation continue well after the wound has been inflicted. Dead tissue left by extensive pulping produces an ideal breeding ground for bacteria. In addition, the action of cavitation creates a partial vacuum in the wound cavity, into which is sucked dirt, clothing and other debris. This increases the infection, encouraging the development of gas gangrene – an often fatal condition. Consequently, treatment is more difficult for the surgeon, and more painful for the patient.

Further complications are caused by the action of secondary missiles. When the bullet enters the body, particles of tissue, notably bone fragments, can be flung outwards from the path of the projectile in a radial fashion, thereby increasing the overall wound area.

Of all target areas the sniper prefers to aim for the head. Although relatively small, the head contains many incapacitating elements, of which the most important is, of course, the brain. Although it is quite possible for an individual to survive surprisingly massive damage to the brain, as a result, for example, of blows from a hammer or axe, the consequences of the hydraulic shock wave and cavitation from high-velocity missiles are devastating. This is especially so because the hard, enclosing presence of the skull prevents the full expansion and contraction of the brain which otherwise might lessen the damage. Typically, a high-velocity bullet will shatter the skull like an eggshell, blasting fragments of bone and brains out of a large exit hole. (This phenomenon can be discerned in newsreel footage of the assassination of President John F. Kennedy in November 1963: a pink mist – blood-spattered brain – is momentarily visible behind his head.)

The nature and consequences of a head wound were graphically described by a medic of the US 30th Infantry Division, fighting in north-west Europe at the end of 1944:

> The sniper's finger presses the trigger and the bullet passes through the helmet, scalp, skull, small blood vessels' membrane into the soft sponginess of the brain substance in the occipital lobe of the cerebral hemisphere. Then you're either paralysed or you're blind or you can't smell anything or your memory is gone or you can't talk or you're bleeding – or you're dead. If a medic picks you up quickly enough, there's a surgeon who can pick out the bullet, tie up the blood vessels, cover up the hole in your head with a tantalum steel plate, then slowly you learn things all over again . . . But if the bullet rips through your medulla region in the back of your head (about twice the size of your thumb) or if it tears through a big blood vessel in the brain – then you're dead buddy.[8]

At longer ranges the sniper aims for the torso. Although not so vital as the head, it is nevertheless a target-rich environment, considerably larger in size and so easier to hit. The abdomen is particularly vulnerable to high-velocity wounds. The effect of cavitation is very pronounced on organs such as the liver, kidneys, spleen and pancreas, which can be damaged even when not directly struck by the bullet. Direct contact is invariably catastrophic. Additionally, the pulsating action of cavitation leads to violent expansion and contraction of the gases in the intestines, producing rupture from within, as well as breaking mesenteric blood vessels to cause severe haemorrhaging. By contrast, the thorax (chest) is mainly an air-filled cavity, and while the heart and great vessels that supply it are highly susceptible to cavitation, the lungs are largely resistant to this phenomenon, although the fracture of surrounding ribs can damage the lungs and produce serious haemorrhaging.

High-velocity wounds to the limbs are necessarily less severe, but can still inflict extensive damage, especially if the major long bones are shattered. Leg wounds tend to be more serious and incapacitating, in part because of the larger blood vessels and because of the effect of the weight of the body bearing down on the wounded leg. A survey[9] based on anecdotal evidence of men wounded by AK–47 rounds suggests that limb wounds remove soldiers from the offensive battle but allow them a limited degree of personal defensive action.

The level of injury is also determined by the bullet itself: its calibre, weight, velocity and manufacture. Since the Second World War, the 7.62mm full-power round has become standard for military sniping, although in certain situations, where the danger of over-penetration is considered relevant, 5.56mm ammunition is used.

As a result of studies conducted during the Vietnam War, the 5.56×45mm M193 bullet fired through the M16 assault rifle was found to possess impressive wounding properties. The bullet's powers were swiftly exaggerated – stories of Viet Cong being killed by hydraulic shock wave following a wound to a finger were not uncommon – but it remains a powerful anti-personal agent, although it should be emphasized that its high wounding potential only functions at short ranges of 150–200 yards or less. While most infantry combat takes place within this distance, the sniper usually operates at greater ranges, where the M193 bullet (or its M16A2 successor, the M855) has only very average wounding potential.

Tests conducted by firing an M193 bullet through a gelatin block at short range provide an explanation for its wounding potential.[10] Before it has travelled four inches the light, 55-grain

M193 bullet starts to tumble, by up to 90 degrees, before breaking into several fragments. This action dramatically retards the bullet's progress, absorbing much of its energy to produce extensive cavitation. More significantly, the break-up of the bullet causes extensive damage as fragments tear through muscle and vital organs.

The far more powerful 7.62×51mm M80 bullet used in many NATO rifles exhibits very different wound characteristics. After penetrating about six inches the 150-grain bullet begins to yaw slowly through 180 degrees to travel base-forward for a distance of up to twenty-five inches. As the bullet begins to tumble it gives off massive energy to produce very extensive cavitation. The bullet does not break up. Although a successful military round, the M80 is too stable in wound-ballistic terms. According to one estimate, the average wound track is slightly less than six inches in length,[11] but the full power of the M80 bullet will only be transferred in a much longer than average wound track. Thus, for example, a bullet which hits a fleshy part of the arm or leg – or even certain parts of the torso, such as a shoulder – may travel straight through the body, failing to release the majority of its energy (assuming the bullet is not deflected by bone or other heavier tissue).

Not all ammunition is so stable. The German-manufactured 7.62mm (IWK 19–65) NATO round (much favoured for its accuracy) has a bullet with a copper-gilded steel jacket which is significantly more brittle and thinner than the copper alloy jacket of the M80 bullet. After travelling approximately three inches the bullet begins to yaw and then fragment, swiftly releasing its tremendous wounding potential.[12] The US .308 Federal Match Sierra bullet (used in the military sniper's M852 round, now replacing the M118) begins to tumble after only two inches and then breaks up to carve a large permanent wound track, before dividing into two separate channels of twenty-two and sixteen inches in length.[13] The conclusion from such tests is that the sniper should ensure he has access to match-grade (or other premium) ammunition with unstable bullets.

*

This survey has examined wound ballistics from the objective viewpoints of the sniper and the surgeon who has to deal with the damage inflicted by a high-velocity bullet. But what of the victim? What does it feel like to be hit by a bullet? Subjective sensory experience is impossible to quantify and so it is difficult to come to firm conclusions.

Men who have survived high-velocity bullet wounds often talk of being struck an enormous blow, as if hit with a baseball or cricket bat. One American soldier wounded in the stomach by an AK–47 round in Vietnam recorded that he felt as if the air had been sucked out of his lungs and breathing was painful and difficult.[14] A British soldier, shot in the shoulder by a Japanese sniper from short range, felt a burning numbness, and another man, hit by a German sniper in the head, noted that 'there was really no sensation except that of receiving a stunning blow, a purplish cloud before the eyes and then blackness, oblivion'.[15]

One immediate consequence of a non-fatal bullet wound is shock, a general term used to describe the debilitating effect of being struck such a heavy blow. (In the stricter medical sense, shock refers to the loss of body fluids resulting from the bullet wound. Inadequate fluid levels may result in the loss of kidney function, and eventually heart failure; a shock victim should receive intravenous fluids without delay.) Closely allied to shock is the experience of pain. This varies from person to person, although evidence suggests that severe wounds to the torso mercifully tend to cut out pain – at least in the short term. By contrast, victims with brain wounds usually suffer intense pain without any delay.

Captain C. Shore described some of the reactions to bullet wounds from British soldiers he interviewed during the Second World War:

> Men lightly wounded by rifle or machine-gun bullets have, without exception, told me that they felt severe pain immediately, but were not knocked down, nor did they faint. On the other hand, severely wounded men have said that they were not conscious of pain, but felt the ground coming up to them as they dropped. One man who had done a great deal of boxing in civilian life and who was severely wounded in the lung, said

that his experience was matched only by a bout in the ring some years before when he received a smashing blow beneath the heart and went down on the canvas with a thud. He felt no pain at all, nor did he lose consciousness.[16]

A more recent account comes from the Falklands War of 1982. Corporal Graham Heaton of 3 Para was involved in the night attack against Argentinian positions on Mount Longdon. In the middle of the assault, Corporal Heaton was severely wounded in the right leg (which eventually had to be amputated). He described the incident to the historian Martin Middlebrook:

> Everything goes into slow motion when you are hit. I saw the tracer coming towards me from my right. It seemed like five minutes before it hit me, but that was the slow motion. I was thinking about moving to avoid it but I didn't have time. Then the rounds hit me and knocked me on my face. It felt just like someone sticking a red-hot poker into my leg. I lay on my face and tried to call to the other blokes but it had all gone cold then and I could only move my left leg. There was no pain; the right leg was really cold, just as though it had been stuck in ice for weeks.[17]

The contradictory nature of Corporal Heaton's sensations, one second hot the next cold, are typical of bullet wounds. The intensity of the experience is so great that words must inevitably fail in describing the feeling of being severely wounded. Certainly, however, we do know from the extensive damage caused by high-velocity bullets that the sniper and his rifle are one of the most deadly weapons systems on the battlefield.

CHAPTER NINE

THE WAR IN VIETNAM

A two-man US Marine sniper team waited patiently in the long grass. They were eight hundred yards above a narrow track, high in the mountains near the Laotian border and the DMZ (demilitarized zone) which divided North from South Vietnam. The track was one of the myriad branches that made up the Ho Chi Minh Trail, and the Marines were confident of intercepting the enemy. Their confidence was justified, as a Viet Cong scout was seen waving forward a thirty-strong patrol. Supported by an eight-man security element from the 3rd Recon. Battalion, the Marine snipers felt able to take on the VC unit, despite the numerical disparity between the two forces. Surprise and skill would make up the difference.

The first bullet from the Marine sniper smashed into the VC scout's chest; a second VC dropped as the shot from the Marine spotter hit home. Unaware of the source of the fire, the patrol froze in astonishment, and then panicked as the Marines calmly emptied their magazines into their confused ranks. In under fifteen minutes, eleven of the Viet Cong lay dead with a further six wounded. As the remains of the communist patrol scattered, the Marines quietly withdrew to safety.[1]

Incidents such as this underlined the effectiveness of sniping in guerrilla warfare. That the US armed forces were able to develop a well-trained and equipped sniper force in Vietnam was a tribute to a small group of individuals who 'kept the faith' following the general run-down in sniping at the end of the Korean War in 1953. Concentrated around marksmanship training units and shooting teams, they also maintained a spirited offensive in the military journals.[2]

These sniper enthusiasts – heretics to many senior officers –

argued that the armed forces must develop a coherent sniper doctrine, instigate sniper schools to teach the discipline as widely as possible, and place the position of sniper on a secure footing, with its own TO/E (Table of Organization and Equipment) to prevent the sniper from being swallowed up by the military machine. Alongside these proposals, there were calls for a new sniper rifle and telescopic sights to replace the M1C/Ds still in service, and for the proper supply of match-grade ammunition. As one author complained: 'Target-match-grade ammunition is made for issue to Army rifle teams. Why can't we supply our snipers with ammunition as good or better?'[3]

A Marine officer, Lieutenant (later Major) Jim Land, took the first positive steps to turn theory into practice when, in charge of the Hawaii Marine shooting team, he set up a scout-sniper school in 1960. In a document outlining his intentions[4] Land noted the dearth of reference material on sniping, although he praised the old First World War classics by McBride and Armstrong, which he still found relevant. He emphasized the mental as well as physical qualities required of the sniper, and recommended that the old M1C rifles be replaced by Winchester Model 70s. These were used by Marines for target shooting and recreational hunting, and there were sufficient quantities in the armouries to equip the Marine snipers, Land's sniper course took two weeks; the first week covered marksmanship, the second fieldcraft and related skills. The school in Hawaii remained an exception – until rising casualties in Vietnam helped provoke a radical change in official attitudes towards sniping.

During the late 1950s the old M1 rifle was phased out by the M14, a modified M1 rechambered to take the NATO standard 7.62×51mm round (a change from the former .30–06 of the M1 and M1903 to .308 calibre). A robust and well-made weapon the M14 was something of a heavyweight, coming in at 11.24lb fully loaded; and its full-automatic fire mode was largely wasted by the soldier's inability to control the rifle. The M14's life as the main service rifle of the US Army and Marines was not to last long, however. Even as the M14 was entering service, experiments were being conducted to develop a new weapon, using a lightweight, reduced-calibre round. The result was the M16

133

assault rifle, firing a 5.56mm (.223in) bullet with a genuinely effective full-automatic capability. Intended for short ranges of around two hundred yards or less (where the vast majority of infantry combat takes place) the M16 was an effective rifle, once initial teething troubles were overcome. Introduced into Vietnam in 1966–67, the M16 has continued as the basic service rifle for US armed forces in its M16A1 and M16A2 variants.

This re-equipping of the armed forces was matched by a transformation in rifle instruction. Traditionally the US military had trained its recruits on conventional rifle ranges in the basics of aimed fire at targets of a given range. Studies conducted after the Korean War criticized this approach as being too remote from battlefield conditions. Consequently, the Trainfire system was adopted in the mid-1950s; this emphasized cover and conceal-ment, and snap-shooting at pop-up targets at differing and usually unknown ranges. Although commendable in attempting to replicate the situations actually encountered in combat, Train-fire failed to instil the essential rudiments of good marksmanship, and shooting standards declined.

In the 1960s the trend towards the application of mass firepower was furthered by the instigation of the Quick Kill programme, designed to exploit the automatic potential of the M16. Quick Kill encouraged troops to spray short-range targets with rapid automatic fire. The rifle was fired in the general direction of the target on the assumption that some bullets must hit the target. Experience, however, proved that this was rarely the case, despite the most prodigal expenditure of ammunition.

Some idea of the increase in the number of rounds fired for each enemy casualty caused, can be observed in studies of US combat performance in this century. During the First World War it was estimated that for each enemy casualty an average of 7000 rounds were fired; a figure which increased to 25,000 in the Second World War. Evidence of the waste of ammunition caused by automatic weapons combined with a careless tactical doctrine was emphasized by the figures for Vietnam, which rose to estimates of over 50,000 rounds fired for each enemy casualty.[5] Given such a poor ratio the need for trained marksmen who could actually hit the target they had aimed for, especially at medium and long ranges, became ever more important. It was a proud boast of American snipers in Vietnam that their ratio of rounds fired to kills secured was equally extraordinary in reverse, just 1.7 to 1. Despite the many difficulties of terrain for long-range shooting, Vietnam would come to be a proving-ground for sniping.

SNIPERS FROM THE NORTH

On 11 December 1961 direct military aid to South Vietnam was sanctioned by the United States government. This aid consisted mostly of the deployment of military advisers and aviation units, acting to support the Army of the Republic of Vietnam (ARVN). By the end of 1964, however, the military situation had deteriorated rapidly, and the following year American ground forces were built up to stabilize the situation. The influx of US troops reached a peak of nearly 540,000 men by 1969; thereafter there was a steady reduction as 'Vietnamization' took effect.

For the majority of American soldiers, Vietnam was a bewildering battleground. The enemy consisted of the elusive and locally based Viet Cong guerrillas, reinforced by regular troops from the North Vietnamese Army (NVA) – many of them combat-hardened by years of war against their former colonial masters, the French. The cultural gap between the Vietnamese people and the Americans was never properly bridged – a major failing in guerrilla warfare – and the extremes of climate and rugged terrain made soldiering for Western armies particularly difficult.

While the US military employed its mastery of technology in an attempt to tip the scales of war in its favour, for the ordinary GI Vietnam became a series of arduous and dangerous foot patrols through jungle and mountain. There they faced the multiple threats of booby-traps, ambushes and constant sniping from the communist enemy.

As in so many other low-intensity conflicts, 'sniping' referred to almost any incoming rifle fire, much of it merely opportunistic sharpshooter forays from VC guerrillas, intended as much to wear down morale as inflict casualties. But alongside the VC, equipped with short-range SKS and AK–47 rifles, were trained communist snipers armed with the 7.62mm M1891/30 Mosin-Nagant sniper rifles (often known in US defence summaries as the K–44). Attached to the M91/30 was either a 4-power PE telescopic sight or the lighter 3.5-power PU. This combination had been the armament of the Red Army snipers of the Second World War, and as it had been effective for them, so it was for the snipers of the NVA.

NVA snipers were all volunteer recruits who were given a three-month training course.[6] For the first two months they concentrated on basic rifle handling, and in the final month they covered the finer points of marksmanship with the telescopic sight. Range firing was conducted against man-sized paper targets at distances of up to a thousand yards, although in the field engagements were not normally initiated at ranges over five hundred yards. Once in combat in South Vietnam the snipers would have little recourse to the services of the armourer, so each sniper was given basic training in rifle maintenance and repair. Further instruction was given to the recruits in camouflage, concealed movement and the selection of suitable fire positions. Snipers were considered an élite within the NVA and morale was correspondingly high.

When their training was complete the snipers were sent down the Ho Chi Minh Trail to infiltrate South Vietnam. The largest organizational unit was the sniper company, consisting of a headquarters element and three platoons. The platoon was the main tactical unit; while deployed in support of NVA main force units it operated in an independent capacity. Within each platoon were three squads of about ten men, which were further divided into cells, usually of three snipers each. The NVA snipers were assigned a local VC unit, to provide guides, reconnaissance agents and security guards. Once in the field the snipers gave the VC elementary training in sniper tactics, and in return received instruction in setting booby traps and ambushes suitable to the local situation. In order to battle-harden newly arrived snipers, it was customary to deploy them against soft targets from the South Vietnamese Army (ARVN) before sending them in to face US troops.

The ambush was the most highly regarded communist tactic, and one in which snipers were able to play an important part. The lead element of the American or ARVN patrol would be shot at by one sniper with the double intention of slowing its advance and drawing it towards his position. The other members of the NVA sniper cell would be positioned on the patrol's flanks and rear, and would begin to fire on the unit commander, radio operator and heavy weapons man. No more than five shots would be fired before the snipers withdrew to new positions. Mines and other booby traps would be positioned in the vicinity to increase the patrol's casualties and allow time for the snipers to withdraw.

NVA snipers were instructed to shoot at helicopters coming in to land and, if possible, eliminate the Command Post helicopter. Using this tactic they achieved a number of successes in the early stages of the American involvement, especially when helicopters would slowly reconnoitre a landing zone – a vulnerable manoeuvre subsequently reduced to a minimum. The introduction of light observation helicopters (LOH) provided accurate firepower which made it dangerous for the communist snipers to operate in the open. An American officer explained: '. . . every time a Viet Cong fired his weapon, unless he was in a bunkered position he signed his own death warrant because the LOHs picked him right off'.[7]

Another favoured communist ploy involved snipers creeping up behind patrols, picking off the rear man, and then immediately slipping away. This procedure would be repeated every day or so, causing slow attrition and eroding morale. To prevent this, US troops introduced 'stay-behind' teams. Using natural cover they hid themselves along the trail at the rear of the main patrol in an attempt to counter-ambush the communist snipers. This left the 'stay-behind' team vulnerable to attack, however, and was only ever a partial solution to the problem.

NVA snipers developed the technique of working their way between two separate enemy units under cover of darkness, firing at one and then the other, in the hope of initiating a fratricidal firefight between the two units. This tactic was witnessed by American Special Forces officer John L. Plaster:

> A lone communist rifleman slipped between our compound in Vietnam's Central Highlands and a tiny [South] Vietnamese Army guard post at a bridge a mile away. Exploiting the natural confusion of darkness the rifleman fired a couple of rounds at the guard post then turned his rifle towards our perimeter and let loose a few shots. Sure as hell, one of our indigenous security men shot back, and since the distant bridge was directly in line, his bullets caused our allies to start throwing rounds our way. It escalated in seconds to a full-scale shootout, under cover of which the anonymous rifleman slipped safely away. For ten minutes we exchanged fire, and only the great range kept anyone from being seriously hurt.[8]

The effectiveness of a good NVA sniper extended beyond securing kills and lowering morale; at times he would be able to hold down large numbers of troops, rendering them tactically immobile for hours on end. One such instance was observed by a Marine sniper, Joseph T. Ward, who described how a communist sniper had pinned down a neighbouring company of Marines. Although an air strike had been called, low cloud prevented the three-strong flight of F–4 Phantoms from attacking for over two hours. During this time, whenever a Marine moved he was picked off by the sniper. Even after the first strike of napalm, the sniper continued his work, wounding another man before departing the scene as a second strike went in. Obliged to admire the skill of the persistent sniper, Ward recalled: 'While we cleaned our rifles, I thought what a day's worth one of Uncle Ho's best had given. One enemy sniper had killed three grunts, wounded four, tied up two companies and six fighter-bombers a good part of the day, and we hadn't seen any sign of him.'[9]

At the height of the battle of the Ia Drang Valley in 1965, when the 1st Air Cavalry Division came up against large numbers of NVA regulars for the first time, snipers were a chronic problem as the troops tried to secure LZs (landing zones) in hostile territory. One report noted how an NVA sniper had infiltrated into the LZ held by troops of the 1st Battalion, 7th Cavalry. Taking aim from a well-concealed position the sniper hit and severely wounded the commander of A Company. 'Even after the enemy sniper had fired,' the report noted, 'he remained undiscovered. The damage to the morale of the unit was severe. They no longer felt secure even within their own perimeter.'[10]

During the siege of Khe Sanh in early 1968, the defending Marines regularly came under sniper fire from the surrounding enemy forces. The quality of this fire varied according to the snipers involved; over time some of them began to take on individual personalities for the Marines. One of the more celebrated was dubbed 'Luke the Gook'; he maintained a constant, if generally ineffective, fire from a hole just over two hundred yards from the Marine lines.

The American reaction to VC snipers became increasingly sophisticated as the siege progressed. Both sides began to adopt the live-and-let-live system (made prominent during the First World War, *see* page 54): the sniper fires a few shots in the general direction of the enemy, thereby discharging his duty to his own side but not inviting the massive retaliation that would follow from accurate sniping. Some of these aspects are contained in an account of fighting at Khe Sanh written by William H. Dabney, then a captain commanding a company in the 3rd Battalion, 26th Marines. Dabney had been ordered to hold a position exposed to sniper fire:

> The sniper was well concealed on a hill about four hundred yards to the north, the only ground high enough and close enough to our positions on Hill 881 South to offer a vantage point for effective rifle fire. He had been there for about a week. He fired rarely but he was deadly. With a total of perhaps twenty rounds, he had killed two of my Marines, and wounded half a dozen others. Even a napalm storm failed to silence him. He had my stretcher bearers pinned down at a time when I had some serious medevacing to be getting on with.
>
> He was careful, but not quite careful enough. On a still afternoon, a machine-gunner spotted a slight movement in a bush. A recoilless rifle, our primary mid-range anti-tank gun, was zeroed in on his spider hole and a 106mm high-explosive

plastic round was sent crashing through the bush. The NVA nest became a crater, and its former occupant a formless pulp.

Another sniper sprang up from nowhere to take his place. He too got a 106. The sniping continued for ten days – this time from a different part of the hill. Again the crew wrestled their 106 around the rough slopes. When the gunner and spotter picked their mark, a young private crawled towards me. Crouched in his foxhole, he had been watching this particular sniper for the past week. The guy had fired about as many rounds as his predecessors, but hadn't hit a damned thing. The private suggested we leave him be for the time being. If we blew him away, the North Vietnamese might replace him with someone who shot straight. This idea made sense and the 106 was moved back. My men even started waving 'Maggie's drawers' at him – a red cloth that we used to signal a miss on the rifle range. Then we figured he might be faking us and we quit the taunts. He stayed there for the whole of the battle – about two months – fired regularly, and never hit a man.[11]

Dabney and his fellow Marines' experience of the live-and-let-live system was echoed by others. As a young officer in the Marine Corps, during the early stages of the American involvement in 1965, the writer Philip Caputo witnessed the phenomenon. Stationed on Hill 268, Caputo and his fellow Marines were not troubled by enemy activity, except for 'sixteen-hundred Charlie, a punctual guerrilla who cranked off a few rounds at four o'clock almost every afternoon. We grew rather fond of him, mainly because he never hit anything.'[12]

THE MARINE CORPS RESPONSE

The Marines were the first American troops to develop sniping in Vietnam. As soon as they had experienced the nature of the guerrilla conflict against the Viet Cong they began experimenting with snipers to take the long-range shooting war directly to the enemy. As early as October 1965, Marine snipers were beginning

to operate on a regular basis; on 23 November one team at Phu
Bai killed two VC and wounded a third at a distance estimated to
be over 1100 yards.[13]

Although a few ex-army M1Ds were used by the Marines, the
main sniper rifle was the Winchester Model 70. Large numbers
of Winchesters had been acquired by the Marine Corps during
the 1950s, and it was only natural that these highly accurate
weapons should be used by Marine snipers. Objections that
match-grade cartridges would cause problems with ammunition
supply were at last overruled. Unertl 8-power scopes were
attached to the Winchester to provide a weapon system capable
of genuine long-range shooting.

Despite its accuracy, the Winchester Model 70/Unertl system
had a number of weaknesses, and in December 1965 tests were
conducted to find a replacement. The main criticism of the Model
70 was that it was a target rather than a true military sniper rifle.
A test survey[14] noted that it was somewhat difficult to bed the
receiver into the stock, and that the trigger was non-adjustable.
More significantly, the Model 70 was chambered for the old
.30–06 cartridge and not the now standard 7.62mm round. The
report continued: 'Modification to the 7.62mm cartridge could be
accomplished, but would be expensive and not especially desir-
able. Additionally, rifles that are nominally "Winchester Model
70" are now found, due to factory modifications made over the
years, in three different types. Many of these parts are not
interchangeable.'

Using the Winchester Model 70 as a reference base, a number
of other rifles were evaluated: the Remington Model 600, Remington Model 700-ADL/BDL, Harrington and Richardson Ultra-Rifle
and the Remington 700-40. Of these the Remington Model 700-40
came out as the winner; not only was it chambered for the
7.62mm cartridge, it was very accurate and deemed sufficiently
robust for military action. Other advantages included a trigger
that was internally adjustable and a relative simplicity in bedding
the rifle.

The long Unertl scope was considered cumbersome, and was
criticized in the report as having unsatisfactory eye relief relative
to the receiver mounting. It also suffered from inadequate seal-

ing, and in the humid climate of Vietnam, where moisture could easily get inside the sight, this was a significant failing. Several commercially manufactured sights were tested and the Redfield AccuRange 3×–9× was selected as the most suitable: it was well sealed, possessed adequate internal adjustment and was fitted with a built-in range finder up to six hundred yards. The Redfield was a variable-power scope, which enabled the sniper to change the magnification, from 3-power, useful in general target acquisition and during poor light conditions, up to 9-power, to give good magnification at long ranges.

The Remington Model 700-40 combined well with the Redfield AccuRange scope to make a good sniper rifle. The first Remingtons arrived in Vietnam during the spring of 1967, and from then on slowly replaced the Winchester/Unertl combination. Subsequently, the Model 700-40 was designated the M40 sniper rifle, which was later upgraded into the M40A1 – the current Marine sniper rifle.

The first sniping school in Vietnam was set up in the 3rd Marine Division. Its commander was Captain (later Major) Robert A. Russell, a veteran with over two decades of experience in rifle team shooting. Russell considered this the perfect assignment: 'For years I've been stressing the fact that the Marine Corps competition-in-arms program has the objective of making a Marine more proficient in combat.'[15]

Given a free hand to organize the school as he saw fit, Russell set about scouring the Marine Corps for top marksmen with experience in rifle instruction. Typical of the five senior NCOs chosen as staff instructors was Gunnery Sergeant Marvin C. Lange, a thirty-five-year-old Texan (a significantly large number of snipers came from the South) with a Bronze Star Medal (with Combat 'V') and a distinguished rifleman's badge. A three times winner of the San Diego Team Trophy, he had helped win the Lloyd Team Trophy at the National Matches.

Russell and his instructors set about organizing the course and producing a training syllabus. But as none of them had yet had the opportunity to conduct sniping operations in the field, they put themselves through the course, and then went out to put theory into practice with the acid test – hunting Viet Cong.

Pairing up into three teams they swiftly experienced the frustrations and rewards of sniping: on one occasion they waited immobile for thirteen hours without a sighting, on another they eliminated two VC parading themselves in front of American positions, beyond the range of M14s and unaware of the presence of Winchester-armed Marine snipers.

Once the sniper instructors had satisfied themselves that they were on the right track, Russell put out a request for volunteers. All sniper recruits had to be recommended by their COs, and be a proven combat veteran as well as an expert rifleman with a minimum recorded score of 220×250. For those men who met the qualifications, sniping offered a chance to escape the monotony of being a grunt ground-pounder and become part of an élite force which was taking the war to the enemy. As one sniper said: 'It's a challenge to sneak up on old Charlie in his own back yard and put a hurtin' on him. I like being a sniper.'[16]

When a suitably long shooting range had been bulldozed out of the scrub, to give ranges varying from four hundred to a thousand yards, the recruits were set to work. Because of the need to get snipers into action the initial courses were only three days long, concentrating on improving marksmanship and gaining familiarity in shooting with telescopic sights. Those who passed the crash course were issued with Winchester/Unertl rifles and sent back into the bush.

After the initial demand for snipers had been met, the course was expanded in length to two and later four weeks. The volunteers were taught the full range of sniper skills, including artillery observation and direction. This was particularly important as snipers were often in good positions to call down artillery or air strikes when large guerrilla forces were located. In addition to their duties at the school, the instructors travelled around the units of the 3rd Marine Division to help set up sniper platoons.

While the 3rd Marine Division was the first to get a sniper school up and running, it was soon followed by the 1st Marine Division. Now a captain, Jim Land was instructed by the commander of the 1st Marine Division, Major-General Herman Nickerson Jnr, to 'train the best snipers in the Marine Corps'. Once in Vietnam, Land set about his task; like Russell he had complete freedom of operation, his only problem being a shortage of equipment and instructors – many of them already taken by Russell's school.

Land was particularly fortunate, however, in being able to secure the services of Gunnery Sergeant Carlos N. Hathcock, an old friend and winner of the 1965 Wimbledon Cup and 1000-Yard Championships. Hathcock was to become a legend in Marine and sniper circles. One of his many exploits occurred during March 1967 when he and his spotter pinned down a company sized NVA unit in Elephant Valley for five days. On another occasion Hathcock stalked his way alone into the headquarters area of an NVA division and killed the enemy commander. As well as amassing a score of ninety-three confirmed kills against the VC/NVA (plus many more unconfirmed) he played a key role in developing Marine sniping in the aftermath of the Vietnam War. The Marine Corps's eminence in modern sniping owes much to Hathcock's influence.[17]

Once Land had gathered together his instructing staff he set up ranges and classrooms on Hill 55, south-west of Da Nang. Land and his instructors (like Russell and his men) went out on combat missions before they took on any students; only Hathcock had prior sniping experience. Crawling through jungles and paddy fields the Marine instructors gained the necessary training to eliminate VC with long-range rifle fire. In an initial three-month period Land's seventeen-strong team of instructors killed more enemy than any single battalion in I Corps.[18]

Land was particularly interested in the motivation of the sniper, and he operated a careful screening process to select only the recruits he considered had the right psychological make-up to become good snipers. While Land conducted the standard interview with potential students, he got his instructors to find out more by taking them to the Enlisted Men's club for an informal drink. Land wrote at some length of these psychological factors:

It takes a special kind of courage to be alone – to be alone with your fears, to be alone with your doubts. There is no one from whom you can draw strength, except yourself. This courage is not the often seen, superficial brand, stimulated by the flow of adrenaline. And neither is it the courage that comes from the fear that others might think you are a coward.

For the sniper, there is no hate of the enemy, only respect of him or her as a quarry. Psychologically, the only motive that will sustain the sniper is knowing he is doing a necessary job and having the confidence that he is the best person to do it. On the battlefield, hate will destroy any man – especially a sniper. Killing for revenge will ultimately twist his mind.

When you look through that scope, the first thing you see is the eyes. There is a lot of difference between shooting at a shadow, shooting at an outline, and shooting at a pair of eyes. It is amazing when you put that scope on somebody, the first thing that pops out at you is the eyes. Many men can't do it at that point.[19]

The sniper had also to be physically and psychologically tough. Operating in the hostile jungles and mountains of Vietnam placed them under great strain, which some were unable to endure. The best snipers were those able to combine their professional and tactical skills with dogged will-power. A Marine captain summed up this requirement: 'You have to be strong enough to endure lying in the weeds day after day, letting the bugs crawl all over you and bite you, letting the sun cook you and rain boil you, shitting and pissing in your pants. Lying there because you know that Charlie's coming and you're gonna kill him.'[20]

As the war progressed so the need for snipers grew and training was expanded; recruits straight from boot camp were accepted on courses as long as they had good rifle scores and the right aptitude. A sniper school had been set up at Camp Pendleton, California, and volunteers were put through a rigorous instructional regime, which also reflected the growing sophistication of sniper training.[21]

The recruits were equipped with Remington 700–40s and given intensive instruction on the ranges. They were expected to fire 120 rounds a day, six days a week, and required to hit a man-sized moving target at 500 yards, a stationary one at 1000 yards.

There was little latitude for error; only a few misses led to failure and dismissal from the course. Alongside shooting, the recruits were taught the other standard elements of sniper training: map reading, judging distance, observation and, in view of the special conditions encountered in Vietnam, survival training. Instruction in directing artillery and air strikes caused some problems, as the snipers worked in feet and yards, while the rest of the armed forces had gone over to the metric system. Consequently, they had to learn to accommodate both systems.

In order to simulate the Vietnamese environment in Camp Pendleton, a mock village was constructed, which the recruits used for a number of exercises, especially to test their scouting skills. When they arrived in Vietnam the sniper recruits would have as good an idea of what to expect as any training camp could provide. Once in the field, the sniper had periodically to 'requalify' by demonstrating his shooting skills on rifle ranges at Da Nang. Failure to achieve a required score could mean reassignment as a grunt.

The success of the Marine sniping programme led to a formalization of the sniper's position within the Marine Corps structure. In June 1968, divisional orders[22] approved the organization and formation of sniper platoons within each regiment's headquarters company and in the headquarters and service company of the reconnaissance battalion. The infantry regiment sniper platoon consisted of three squads of five two-man teams and a squad leader each, plus a senior NCO, an armourer and an officer, with a total strength of one officer and thirty-five enlisted men. The reconnaissance battalion had four squads of three two-man teams each under a squad leader, plus a senior NCO, armourer and officer, with a total strength of one officer and thirty enlisted men. Combat conditions, however, necessarily modified the standard organizational structure.

As sniper teams would often be operating deep in 'Indian' country, beyond the safety of fire bases and other defended areas, they were usually assigned escorts, their size depending on the depth of penetration. Squads of ten to fourteen Marines were assigned to teams on semi-independent missions, when operating beyond the sight (or radio communication) of parent units; smaller escorts of four to eight Marines were assigned for missions within two thousand metres of friendly troops.

In practice, however, snipers often preferred to work with smaller escorts, or none at all if surprise and stealth were paramount. Such missions took a special type of cool courage for a two-man team, alone against a ruthless enemy who rarely took prisoners except for the purposes of mutilation and torture. Snipers had substantial VC bounties on their heads, which was further evidence of their effectiveness, yet a real danger, as they could be overwhelmed by local villagers eager for financial reward. But at least snipers had the comfort of knowing they were provided with appropriate communications, with artillery and mortar fire support on a constant 'on call' basis.

Within the sniper team it was usual for the team leader to carry the bolt-action rifle while his spotter was armed with an accurized M14. Occasionally M16s were carried by spotters but they lacked long-range accuracy, and as the M14 and Remington shared the same 7.62mm calibre, their ammunition could be shared. The spotter carried a 20-power scout telescope (if available), a useful device for long-range observation and for checking the sniper's fall of shot. The team would always have a pair of binoculars for general observation, and where necessary, various night-vision aids.

The team leader would normally have considerable combat experience while his spotter would act as an apprentice, before being promoted to team leader, if sufficiently able. Soon after his arrival in Vietnam in April 1969, Joseph T. Ward was assigned to act as a spotter to a veteran sniper and top shot called Chuck Mawhinney. As part of the breaking-in process, Mawhinney took him out on a routine squad patrol. Ward recounted his first taste of battlefield action:

Chuck wanted to see how I handled myself. About an hour away from camp, the point man spotted some VC moving through a large clearing nine hundred yards in the distance. The squad crouched low and the word was passed for 'snipers up'. Chuck was already running towards a small rise about seventy-five feet away on the other side of a rice paddy, with me right behind.

The lieutenant was yelling and swearing as loud as he dared for Chuck to slow down and watch out for booby traps. If Chuck heard him, he didn't give any indication, and in a few seconds he assumed a kneeling position. As trained, I took my spot two feet to his left rear. By the time I focused my field glasses and confirmed they had weapons, Chuck had his breathing under control, and the first shot rang out.

A VC at the center of the column dropped. I heard myself say, 'Hit', the same way I had said so many times at the rifle range. It came out of my mouth calmly, well-rehearsed. All the VC, all but one, began running and dragging their fallen comrade towards the tree line they had come from. A second shot broke the early morning quiet and a man towards the front of the column fell. 'Hit,' I said as I watched them pick up the second body and make the safety of the trees. The man that had started for the opposite tree line had made two mistakes. His first was getting separated from the main unit. His last was trying to rejoin them with a sniper watching. Another shot rang out, and he fell behind a rice paddy dike. For the third time in barely half a minute, the word 'Hit' came from my mouth. I felt nothing. The reality of the situation hadn't sunk in.[23]

To be able to kill VC at such ranges was a great boost to the US infantryman's morale; at last the elusive Viet Cong was being hit hard. For the Viet Cong, morale fell to a corresponding degree, now the hunter was being hunted.

Despite the success of Mawhinney and Ward in this instance, snipers often experienced problems working with infantry patrols. The infantry were usually too slow, cumbersome and noisy to give the snipers a good chance of taking the enemy by surprise. Also, the infantry squad commander and the snipers had different priorities, and thick jungle did not exploit the possibilities of a bolt-action rifle. Snipers needed freedom of action in order to operate effectively, and a wise commander allowed them to work in their own way. Fortunately, as a semi-independent entity within the Marine Corps, the snipers were normally able to plan and execute their own missions.

In time, orders were issued to infantry unit commanders explaining how to use sniper teams to greatest effect.[24] The teams were to be assigned fire squads and a radio operator for security; the teams were not to be tied down to conventional infantry operations such as acting as guards or be used in ambushes, patrols or in listening posts at night; and adequate liaison between snipers and infantry was to be maintained.

In turn, the snipers were able to aid the infantry commander, reducing the enemy presence and providing intelligence of VC movements and intentions. The reconnaissance element of Marine scout sniper operations was not forgotten. All snipers kept logs of their operations, which were collated with those of their comrades to form a consolidated report of sightings in the field. Reproduced here is a sniper's log submitted to the headquarters of the 7th Marine Regiment, a document similar in style and scope to the summaries prepared by British snipers during the First World War (see page 70). Because the sniper was highly skilled in observation, and so often out in the field, his log was a significant piece in the intelligence mosaic.

SNIPER LOG

Date: 12 Dec 66
Time: 0800
Place: Hill 55 finger 1 (coord 981625)

1500 – Received sniper rd. at river, could not locate sniper.
1530 – Spotted suspicious man, had him picked up by patrol, neg results.
20 Dec Finger 1
0915 – Spotted smoke coming out of ground in rice paddy.
1100 – Called back to S–2.
21 Dec Finger 1
1330 – Spotted 1 woman with M1 rifle (991627) running thru the tree line during art barrage.
22 Dec Finger 1

0800 – Spotted 6 men and 1 woman in cane field (992620), seemed to be laying mines or booby traps, observed for 4 hrs.

1200 – 1 man picked up M1 rifle, moved south (992618). 2 other men came across rr tracks to some area carrying carbines. Many people into area wearing ponchos. Art mission called on area, results were 3 KIA (confirmed).

1400 – 'A' Co, 1–26 (2 platoons) sweep area and found mines in cane field.

22 Dec 'B' Co

1030 – Spotted 1 VC wearing khaki shirt and believed to be carrying rifle (899551) (350 yds) 1 shot, 1 KIA. (Sgt Hathcock).

1045 – Spotted 1 VC wearing khaki shirt and black cut off trousers and believed to be carrying rifle (892552) 1 shot, 1 KIA, shot with model 70 8-power scope.

1115 – Spotted 1 VC carrying white satchel bag wearing white pyjamas (600 yds) 1 shot, 1 WIA, with model 70 8-power scope (Sgt Hathcock).

1230 – Spotted 2 VC wearing khaki clothes carrying rifles, fired 1 rd. Results, 1 KIA, the other fled.

1240 – Spotted 1 VC in field wearing poncho with butt of rifle protruding through poncho, fired 8 rounds and got 1 WIA (400 yds).[25]

Besides the important work of information gathering, snipers had another secondary function, that of calling in air and artillery strikes. The sniper's knowledge of the local terrain, combined with his map-reading abilities, meant he was well placed to guide ordnance accurately on to the target area. Although snipers preferred to take on the NVA or VC directly through the scopes of their rifles, when enemy numbers were too great it was advisable to call for help.

After nine months in Vietnam, Joseph T. Ward had worked his way to sniper squad leader, based on Hill 65 at An Hoa. Towards the end of his tour of duty, Ward found himself in a situation where, by calling up air and artillery strikes, he was able to inflict massive damage on the enemy. Having been informed of increased enemy activity in the area, Ward set off on a predawn hunter–kill operation, accompanied by his M14-armed spotter, Lance-Corporal Terry Lightfoot.

The two Marine snipers set up a secure position, overlooking a large open stretch of fields and rice paddies where it was likely the enemy would pass. As dawn broke Ward and Lightfoot observed large numbers of NVA – company strength and more – breaking out of the tree line opposite to cross the open ground towards the snipers. Ward contacted the air controller and was told that two F-4 Phantoms armed with napalm were seven minutes from the target area, having just aborted another mission. The F-4s were redirected, but they would have to be swift, otherwise the snipers would be overrun.

As the F-4s closed towards the target, Ward and Lightfoot prepared to take on the advancing communist forces; the finer points of sniping were to be abandoned in favour of a long-range shoot-out. The two Marines hoped to hold the enemy in the rice paddies and along the tree line, to make them an ideal target for the Phantoms. Ward wrote this account of the action:

Lightfoot was laying M14 magazines in front of him so he could get to them more easily. I was heartened to see that he had the instinct to know that we were about to get into an out-and-out shooting match. He'd never tangled with that many enemy troops; neither had I. They could overwhelm us by sheer numbers, but we had the element of surprise, marksmanship and concealment on our side. I looked at my watch with one eye and at Charlie with the other through the rifle scope. They had started moving, and the lead man would be within Light-foot's 700-yard limit in less than sixty seconds.

'Now!' I shouted at Lightfoot. He began firing, and at the same moment I shot a machine-gunner between the eyes. Lightfoot hit the three lead men before they could all take cover. My next shot took out the second machine-gunner. My third hit a mortar man in the neck.

Shot four hit the second mortar man; just as he dropped a round down the tube, he fell over like a rag doll and the result of his last act as a soldier was a misplaced mortar that landed among his own people who Lightfoot had pinned down, killing two of them. They thought it was from us and that we'd hit them with mortars too. When they moved, Lightfoot dropped two more.

My fifth shot hit the ammo man starting to take over the first machine-gun. There were a lot of AK rounds randomly hitting the hill side. They hadn't located us, but if the machine-guns and mortars were allowed to work steadily they would eventually find us.

Lightfoot was doing a number one job of pinning down what was left of the forward squad. They weren't able to get into the fight at all, and I could, for the time being, forget about them.

I looked back at the second gun. It was momentarily out of action. I moved the scope back to the first machine-gun just as it began to belch smoke. I had to get back on the radio to listen for Big Ten [call sign of the lead F-4] in case his final approach was off target. In my haste I missed the man on the gun with my sixth shot.

'Son of a Bitch!' I yelled, as I picked up the handset with one hand and began to reload with the other. 'Come on Big Ten,' I thought, while I chambered a round and put the first of five rounds into the magazine.

Lightfoot was methodically, calmly, firing at the men trapped in the rice paddy. I was glad I wasn't one of them. He wiped the sweat from his eyes and swore once when a bullet hit close enough to kick dirt in his face. Other than that he said nothing.

Charlie had both guns going and was raking the side of our hill with unnerving precision. I watched as tracers richocheted in increasing numbers nearer and nearer to our position and wondered if I had cut it too close.

I'd just put in round number four when I heard the voice of an angel over the radio. 'Long Rifle [Ward's call sign], this is Big Ten. I'm on final approach. Over and out.'

I looked through the scope in the direction I expected Big Ten to be coming from. He was barely visible, but right on the money. Twenty more seconds, that's all we needed. I put in the last round and switched back to the trees and shot a third machine-gunner brought into action.

Charlie was making too much noise trying to find us to hear

Big Ten coming, and the jets were almost on them. I had time for one more shot and took out a mortar man, when the rockets and guns of Big Ten got Charlie's attention. Seconds later the tree line erupted in flames and black smoke. All enemy fire in our direction stopped, the only sounds were from Lightfoot, still working over those poor bastards in the rice field, and the faint roar of napalm devouring trees and humans.[26]

Following the air strike Ward switched his set to the fire control at An Hoa, which swamped the area with a blanket bombardment from several batteries of 105mm and long-range 175mm guns. As the artillery pounded the enemy positions, Ward and Lightfoot edged away from their hide and returned to base. As a result of their sniping prowess, and ability to bring down fire accurately under pressure, the two-man sniper team had severely mauled an entire NVA battalion.

THE ARMY SNIPER PROGRAMME

Faced by VC/NVA snipers and ambushes, the US Army began to use snipers from 1966 onwards and monitored their progress in various study/evaluations. But in contrast to the Marine experience, the Army's sniper experiments produced only mixed results. This disparity had many causes but the fundamental point lay in differing approaches to the organization and employment of snipers.

In detailing the sniper's role, the Marine Corps was quite explicit: 'The mission of the Scout Sniper Platoon is to support the Infantry/Reconnaissance Battalions by providing a specially trained and equipped unit capable of rendering sniper support in combat operations, by providing personnel trained to kill individual enemy soldiers with single rifle shots from positions of concealment.'[27] In other words, the Marine sniper was organized in his own unit and was given freedom of action to use his special skills of long-range shooting.

By contrast, the Army sniper was tacked on to the infantry

platoon or squad. He was not a soldier who had undergone the most rigorous training to gain a coveted position of sniper, but merely an ordinary infantryman who was a good shot and handed a rifle with a telescopic sight. An evaluation conducted for the Combat Development Command during 1966–67 made this note on sniper deployment: 'He remains under the control of his unit [platoon/squad] leader and engages targets of opportunity within the framework of his unit's operations. The mission of the sniper in this role is to extend the effective rifle firing range of his unit.'[28] In this position the Army sniper was hardly a sniper at all, rather a soldier who could shoot further than his fellows. For a true sniper, having the ability to hit targets at long range was only part of the equation; equally important was the mastery of fieldcraft and tactics combined with a freedom of action that brought the enemy within the scope of his rifle.

The Army did not follow the Marines in adopting the bolt-action Winchester and Remington rifles, but preferred to stick with the M14 design. Two M14 models were used; the first an accurized version of the basic model, using specially selected and fitted parts to improve accuracy, and the second, the National Match model, constructed from the outset for target range shooting. Although not as accurate as a bolt-action rifle (the large number of moving parts of an automatic rifle made it less steady) these improved M14s were highly effective out to at least seven hundred metres when using match-grade ammunition. The Redfield AccuRange scope was attached to these M14s.

The failure of the Army snipers to gain results did not go unnoticed and steps were made to remedy the situation. The CDC Trip Report, 'Sniper Programs', issued in April 1969, identified the main problem areas and suggested appropriate solutions.[29] The report noted that snipers were considered as a means of taking on the enemy on the same level as claymore mines and mortar fire, forgetting that snipers were a special asset. 'There is considerable misunderstanding', the report continued, 'in the minds of some officers and most of the men to the contradistinction between ambushing and sniping. The unique capability of the sniper to instil terror through swift, silent, mysterious death is not fully appreciated or capitalized upon.'

A further problem highlighted by the report was the reluctance of many soldiers to engage the enemy, for fear of giving their position away and inviting retaliation – a syndrome familiar since the First World War. Snipers who were part of an infantry squad or platoon were particularly vulnerable to the pressure of their peers; men in dedicated sniper units seldom felt these pressures and delighted in taking the war to the enemy, regardless of the consequences.

In conclusion, the report recognized the need for the Army to produce a doctrine for the employment of snipers, recommending that they be organized primarily as a brigade asset, operating on a semi-independent basis along similar lines to those of the Marine Corps.

Terrain was the other key determinant in the success of sniper deployment. The heavily wooded lowland areas to the north of Saigon, for example, severely hampered sniper activity, although in many parts of South Vietnam snipers could be used to good effect. An area well-suited to sniping was the Mekong Delta, which became the responsibility of the 9th Infantry Division. Under the thoughtful and energetic leadership of Lieutenant-General Julian J. Ewell, the 9th Division initiated a highly successful sniper programme, due in large part to the division's own operational research approach to solving military problems.

Early in 1968 Ewell and the divisional staff began to train men to be snipers, as a result of close consultation with the AMU (Army Marksmanship Unit) at Fort Benning. Major Willis L. Powell and seven NCOs from the AMU arrived in Vietnam in July 1968 to organize the sniper training courses. Progress was slow but by early November the first graduates from the sniper school were ready to go out into the field. The course lasted eighteen days, and training was rigorous, with around 50 per cent of volunteers failing to make the grade. In selecting potential snipers, Army trainers looked for four qualities. First, the recruit should be a good, conscientious soldier, a 'solid citizen'; second, he should be a good team worker; third, he should be a 'daring-type person'; and last, he should be cool under pressure. Although perhaps contradictory requirements they demonstrated the difficulties in finding the right men for the job.

By December 1968 a full complement of seventy-two snipers had emerged ready for action. The divisional staff were hopeful of good results yet were almost immediately disappointed; the end-of-year figures revealed only eight kills for November and eleven for December. In a monograph written by Ewell and his Chief of Staff, Major-General Ira A. Hunt, on the 9th Infantry Division in Vietnam, the issue of sniper deployment was put under the closest scrutiny. The authors wrote:

> We hit upon the flaw in the system, and while the solution was extremely simple, it had an immediate effect. Initially, snipers had been parcelled out by battalions on the basis of two per line company. The company commanders, then, had the responsibility for the snipers and most company commanders could not care less. They used snipers just as any other riflemen. This was the reason we were not getting results. Consequently, we directed assignment of the snipers to the battalion headquarters and held the battalion commanders responsible for their sniper utilization and for their emphasis on the program.[30]

The success of the new system of using snipers in a semi-independent capacity at battalion level was proved by the number of kills made. Throughout the first part of 1969 the figures increased dramatically, reaching a peak in April with 346 enemy killed during the month. Inevitably, the numbers levelled off as the VC became more cautious in the light of US sniper activity, but they still stood at around two hundred per month.[31] An added bonus was the general increase in confidence among the infantry; here was hard evidence that the elusive enemy could be nailed down by skill at arms, and patrols became more aggressive as a result.

As the 9th Infantry Division was deployed in the Delta, certain of its units operated in support of the 'Brown Water Navy', patrolling the great waterways that flowed into the South China Sea. Tango boats, which acted as landing pads for helicopters, were used by sniper teams because their slow speed made them good shooting platforms. The Tango boats would cruise slowly along the Mekong, parallel to the shore, anchoring at regular intervals. The snipers from the 6th Battalion, 31st Infantry, killed thirty-nine Viet Cong during the period 12 April to 9 May 1969 – testimony to the effectiveness of the tactic.

Sergeant Adelbert F. Waldron III was the 9th Division's top sniper, credited with 109 confirmed kills, and he used his expertise to good effect while on riverine patrols. Ewell and Hunt describe an incident involving Waldron: 'One afternoon he was riding along the Mekong River on a Tango boat when an enemy sniper on shore pecked away at the boat. While everyone else on board strained to find the antagonist, who was firing from the shoreline over nine hundred meters away, Sergeant Waldron took up his sniper rifle and picked off the Viet Cong out of the top of a coconut tree with one shot (this from a moving platform).'[32] For his outstanding abilities as a sniper and for bravery in action, Waldron was awarded two Distinguished Service Crosses.

On land, snipers were used in conventional operations, constructing ambush hides overlooking likely enemy trails and meeting places. As the VC were lax about adopting cover at long ranges, snipers were particularly effective in killing the enemy at ranges over six hundred yards. Stay-behind teams comprising two snipers, a radio man and three South Vietnamese Popular Force soldiers were used against VC tax collectors crossing the border from Cambodia and extracting money from the local farmers. Taking up positions just outside a village, the snipers would be directed towards the right targets by local Popular Force soldiers. The levels of VC taxation were swiftly reduced.[33]

In a study of tactical and material innovations used by the US Army during the war in Vietnam, John H. Hay Jnr recorded this incident of Army snipers being used in a counter-sniper role:

At about 0800 hours one morning, a 9th Infantry Division battalion had been conducting a reconnaissance in force when suddenly the men heard the unmistakable sound of rifle fire. Instinctively the point man hit the ground and rolled for cover, but there was only silence. Just one shot had been fired. Somewhere out there was a Viet Cong sniper. The point man surveyed the area; the only possible location for the sniper was in a wood line about 700 meters in front of him. He summoned the radio operator and reported the sniper fire to his platoon leader. A short time later the report reached the battalion commander, who immediately deployed his own sniper team to the point man's location. With his optical equipment the sniper team began a search of the tree line. Finally, the Viet Cong sniper was discovered in a tree 720 meters away. While one team member judged the wind using the M49 [20-power] spotting scope, the other man fired one shot killing the Viet Cong sniper.[34]

The 9th Infantry Division was aided by suitable terrain, but the speed with which it turned around its sniper programme was remarkable. Other divisions began to follow suit but just as Army sniping practices were improving, the ground war was scaled down. As in other conflicts, many of the valuable lessons of sniper deployment would largely be lost in the succeeding years of peace.

TECHNOLOGY AND THE SNIPER

Traditionally snipers have not placed much reliance on technological progress: a good but straightforward bolt-action rifle and human skill was seen as the essential qualification for success. But the expansion of military technology that took place during the Vietnam War – as American strategists went all out to defeat the Viet Cong with every device at their disposal – inevitably had an influence on the conduct of sniping.

Infra-red light-imaging devices, capable of cutting through the veil of darkness, had been used in Korea and even during the latter stages of the Second World War. Cumbersome infra-red sights had been mounted on the M1 carbine (redesignated the M3) but had not been particularly successful. By the early 1960s, however, night-vision devices had greatly improved; they could see further and were far lighter and smaller. As well as infra-red equipment (which has the disadvantage of giving away positions to anyone else with a similar device) a new generation of passive image intensifying devices had come into being. These worked by electronically amplifying the faintest light sources (star or moonlight, for example, as well as infra-red) so that reasonable night vision could be provided to the viewer.

The large tripod-mounted Night Observation Device (NOD) had a range of over a thousand yards, while the smaller AN/PVS-2 Starlight scope was capable of being mounted directly on to a rifle and used as a sight to a distance of about four hundred yards. Although the ghostly green picture produced by an image intensifier had a number of technical shortcomings (as well as weighing over 5lb) it gave the US soldier a distinct edge over his adversary. The Vietnam expression that 'the night belongs to Charlie' may have been largely accurate, but with the introduction of night-vision devices, Charlie's tenure of darkness had become increasingly less secure.

Mounted on the M14 rifle the Starlight scope was of enormous benefit for snipers, and even though ranges were reduced, it extended the operational employment of the sniper into covering night actions. They were also used for certain daytime missions in jungle terrain, where light was poor and distances too short for long-range shooting. The Army snipers of the 9th Infantry Division found night observation aids particularly valuable, and by February 1969 it was recorded that most sniper kills occurred during the hours of darkness. Infra-red searchlights – known as pink light – were used in conjunction with starlight scopes, providing extra illumination during periods of poor ambient light.

The 9th Infantry Division was quick to exploit the possibilities of image intensifiers in night-search helicopter missions. A sniper team would be positioned in a command and control helicopter, its task to act as spotters. When enemy activity was discovered they would fire tracer rounds at the target to direct the fire of waiting helicopter gunships. Over a six-month period 1800 enemy casualties were recorded, which not only disrupted the usual night-time patterns of Viet Cong movement, but greatly affected their morale. Subsequently, the airborne snipers proved effective in hitting the enemy with their own Starlight-scoped M14s, and around 15 per cent of the total kills for night-search operations were attributed to direct sniper fire.

Silencers – or more properly, noise suppressors, as no shot is ever 'silent' – had been employed in military actions since the First World War. Usually suppressors were fitted to pistols or sub-machine-guns for close-range assassination work, and only a few, largely unsuccessful, attempts were made to combine them with rifles. In Vietnam, however, some success was recorded with suppressed sniper rifles, especially when used in conjunction with Starlight scopes during night missions.

A suppressor consists of a tube, fitted over the end of the muzzle, containing a number of baffles which trap the gas coming out of the muzzle and dissipate it slowly into the atmosphere, thereby reducing muzzle noise. A secondary advantage of the suppressor is that it almost entirely eliminates muzzle flash, an important consideration in night operations. Reducing the sound from the rifle's muzzle is only part of the process, however; once the bullet leaves the barrel it creates its own sonic boom – the distinctive crack of the high-velocity bullet. Only by reducing the power of the cartridge, so that its muzzle velocity is under the speed of sound (around 1100 feet per second), can the sonic crack be eliminated. This can be achieved, but accuracy and wounding power of the bullet are greatly reduced as a consequence.

Taking these limitations into account, suppressors could be highly effective in covert operations where the need for stealth outweighed the need for accuracy. Close by, a suppressed shot sounds no louder than an air pistol, and at fifty yards is inaudible. US Special Forces conducting clandestine cross-border missions found suppressors particularly useful. For the ordinary sniper, for whom long-range accuracy was crucial, they were of limited value, although one report notes the bafflement of the enemy when fired on in this manner: 'The VC just seemed to mill around, even after a couple of them had been dropped.'[35]

As in Korea, the .50-calibre M2 Browning machine-gun fitted with a telescopic sight was used by snipers in fixed positions. In many fire bases a twenty-foot-high sand-bagged tower would be built, equipped with an M2 which was able to cover the surrounding country on all sides. As weight was not a problem, Night Observation Devices were directly mounted on the M2, and constant day and night defence maintained.

Whereas the Winchester/Remington and M14 sniper rifles

began to lose accuracy over 800 yards, the specially prepared single-shot .50-calibre M2 was able to hit targets nearly a 1000 yards away in darkness and, in skilful hands, over double that distance in daylight. Joseph Ward remembers sinking an empty sampan with an explosive bullet at a range of 2000 yards,[36] and Carlos Hathcock killed a VC supply carrier at the extraordinary distance of 2500 yards, using an M2 with a Unertl scope.[37] The success of .50-calibre shooting in Vietnam eventually prompted civilian gunsmiths to experiment with single-shot bolt-action weapons, which were of a sniper-rifle quality and yet man-portable.

The interest aroused in sniping as a result of the Vietnam War was not entirely lost after the US withdrawal of its ground forces in 1972. While the military authorities carried out their standard practice of down-grading the sniper presence within the infantry battalions, there remained strong and wide commitment to sniping. Not surprisingly this was centred within the Marine Corps, and over the succeeding decades it was to expand into a new approach to sniping in peacetime that would eventually embrace both the Army and Marines.

CHAPTER TEN

BRITISH SNIPING AFTER 1945

Slowly and haphazardly, sniping in the British armed forces had achieved a respectable standard by 1945. And while standards were not even throughout the Army, there were large numbers of soldiers who had experience of sniping on the battlefield. Unfortunately, this experience was not exploited. Instead of setting up a central school for training and development, sniping was left to the battalions, where it was inevitably pushed aside through pressure of other commitments. None the less, the concept of sniping did not disappear altogether. Attached to the battalion headquarters was a sniping section of eight men under the command of a sergeant, with a further eight men acting as a sniper reserve, there being sufficient equipment for eight two-man sniper teams.[1]

While the British Army's main military effort was devoted towards deterring the Warsaw Pact nations, as part of the commitment to NATO, the bulk of the actual fighting was conducted in small wars across the globe. In what became a withdrawal from Empire, British armed forces were constantly engaged in a series of actions to resolve local crises and ensure an orderly transition of authority to the nominated local power. In many of these bush wars the opportunities for the deployment of snipers was limited, as, for example, in the jungles of Malaya during the 'Emergency', and later in Borneo during the confrontation with Indonesia.

Other theatres of war favoured sniping, especially the Arabian peninsula, where British troops were engaged in several limited wars from the late 1950s onwards. In 1958 the SAS

(Special Air Service) Regiment was called in to support the Sultan of Oman against an Egyptian-backed tribal revolt. The rebels had gained control of the Djebel Akhdar, a vast mountain massif rising to 8,000 feet and considered impregnable. To the men of the SAS, recently flown in from combating communist insurgents in Malaya, fighting a war against hardened tribal warriors in this extreme mountain terrain represented a real challenge. Through a combination of good intelligence, planning and solid determination, the Djebel Akhdar fell to the SAS.

This, and the undercover conflict that followed in Dhofar in the 1970s, were among the last of the 'old-fashioned' wars, reminiscent of the North-West Frontier in the previous century. The terrain was open with long, clear fields of fire, where accurate long-range shooting became a factor in everyday tactical operations. The local tribesmen were found to be good marksmen, and even though they might be armed with rifles dating back to the nineteenth century, such as the Martini–Henry, they were deadly against troops who failed to take appropriate cover. In this environment, British sniping was encouraged to develop.

The Aden Emergency (1963–67) was a confused campaign as Arab guerrilla forces fought the British and each other for control of the territory at the southern tip of the Arabian peninsula. The intensity of the fighting increased steadily as the conflict progressed; sporadic outbursts in the Radfan Mountains were followed by civil insurrection in the city of Aden. The most bitter fighting in Aden was centred around the business quarter, known as Crater, when on 20 June 1967 British troops were ambushed and weapons seized. The British had been committed to maintaining a softly-softly approach, and temporarily withdrew their forces in order to reduce the chance of increased casualties to the civilian population.

As its name suggested, Crater was built within the crater of an extinct volcano and comprised a labyrinth of streets and back alleys, which provided excellent cover for the Arab nationalist guerrillas. To regain control, the British isolated the area and instigated a series of patrols to locate the guerrilla locations and put pressure on them before sending in the main force of a battalion of Argyll and Sutherland Highlanders.

Supporting the Highlanders was the Royal Marines 45 Commando, whose snipers painstakingly clambered up the precipitous crags which surrounded Crater. There they were able to take up positions overlooking the city, and look for enemy activity. For a period of ten days the Commando snipers fired down into the crater; particular satisfaction was gained from killing those terrorists armed with British weapons captured on the 20th. One sniper gave this account of the action:

> The Argylls under Mad Mitch [Lieutenant-Colonel Colin Mitchell, CO of the Argyll and Sutherland Highlanders] made all the news when they retook Crater, but without us they'd never have done it without taking lots of casualties. It was us who wore the Arabs down, and made them realize that they couldn't have it all their own way. We were right up above the town and I could look through my telescopic sight and see them moving about. We waited for hours. The heat was incredible, bouncing off the rocks and burning you to a turn. But it was worth it, when I saw an Arab coming out of a mosque carrying a rifle. I knew he was a terrorist; we'd found out that they used mosques as meeting places. As he walked up this alleyway towards me I took aim. He couldn't see me, of course. And then I squeezed the trigger, and he fell backwards, dropping his rifle. I'd definitely got him. Everyone else around him ran off in panic, leaving him lying there until nightfall.[2]

The Royal Marine snipers steadily reduced the ranks of the terrorists, lowering the morale of the others. Two snipers from 45 Commando fired twenty-five rounds and killed eleven terrorists and wounded a further five. On one occasion a particularly troublesome Arab sniper, well protected in a house on the southern side of Crater, was despatched by a shot from a Carl Gustav anti-tank round. The Commando snipers came under fire, in turn, and two men were wounded, but rather than give away their position they remained in their hides until nightfall, when they were able to crawl away unobserved. The softening up by the Commando snipers made it much easier for the Argylls when they moved into Crater in the early hours on 4 July and recaptured it without a fight. Although the operation eventually amounted to little, as the British left Aden for good in November 1967, it was an instructive example of the effectiveness of sniping in an urban environment.

British troops fought their sniper battles armed with the old Number 4 Mark I(T), which had been the standard sniper rifle during the Second World War. The change of calibre from .303 to the NATO 7.62mm in the 1950s caused problems with ammunition supply. The standard .303 No. 4 infantry rifle was replaced by the 7.62mm L1A1 SLR (Self-Loading Rifle), based on the Belgian FN FAL. Although an excellent service rifle, the SLR's long and lightweight receiver made it a poor sniper weapon, and even though telescopic sights were mounted on some examples, they were never particularly successful.

In keeping with the low priority of sniping within the Army, it was decided not to find a new sniping rifle, but to convert the existing No. 4 to 7.62mm. An earlier attempt to convert the calibre, the L8(T), had been a failure, but a later conversion, known as the L42A1, was a success and remained in British service until the mid-1980s. The L42A1 had a longer and heavier barrel than standard, and its most distinctive feature was the shortening of the fore end and hand guard. Despite its limited magnification (3-power) the old Mk 32 telescopic sight was retained for the L42A1.[3]

The year following the British withdrawal from Aden in 1967 was the only one since the Second World War that British forces were not engaged on active service. In the summer of 1969, the Army was called out by the civil powers in Northern Ireland to restore order between battling elements of Catholics and Protestants, as the Province collapsed into near anarchy. Initially, the Army was engaged in separating the warring factions but by 1970 it was engaged in a fight with the terrorists, in particular the IRA (Irish Republican Army).

The early 1970s was the period of the 'shooting war' in Northern Ireland, where terrorist gunmen directly took on the Army. When moving through the streets of Belfast and Londonderry, British soldiers found themselves vulnerable to IRA gunmen, who were able to fire a few shots at a patrol and then disappear into a rabbit warren of houses and a sympathetic population. Although these IRA activities could hardly be called sniping in the military sense, they were causing the Army casualties. The night of 3 July 1970, for example, started with rioting in Republican areas which developed into a running gun battle, leading to thirteen soldiers receiving gunshot wounds.

This phase of the conflict did not last long, as the Army's tactics became more cautious and flexible, presenting fewer easy targets for the gunmen. The soldiers acclimatized themselves to their new environment and began to turn the tables on the gunmen, who in the end were outshot. Increasingly, the IRA turned towards softer targets, such as the police or off-duty soldiers, and began to make greater use of bombs as the mainstay of their offensive.

The shooting war lasted longer in the countryside, especially in the border areas of the Province which were fiercely Republican, hard to patrol and near the sanctuary of the Irish Republic. Hiding in hedgerows and rough ground the IRA began sniping at the security forces. But they soon found themselves up against Army and Royal Marine snipers, whose training in counter-sniping tactics severely curtailed the IRA initiative. For security reasons there is little hard information on sniper activities in Northern Ireland, although there were several notable sniper actions in the 'bandit' country of South Armagh in the early 1970s. Later in the decade the Army became more concerned with intelligence gathering, both in the towns and countryside. Since then direct military action has largely been left to specialist units such as the SAS, MRF and 14 Intelligence Company.

The IRA announced a return to a sniping campaign, with the shooting of a British soldier in August 1992, as he stood in the exposed market square of Crossmaglen. In November a police constable in Fermanagh was shot from over the border, while in 1993 several men from the security forces were killed in South Armagh from long-range rifle fire. Contrary to former IRA practice, where cowboy gunslinging was the norm, it was clear that the security forces were being hit by a trained sniper (or snipers). One Army source admitted: '. . . you don't get a kill on your first and only shot without a fair bit of practice . . . We are facing a high degree of skill from these terrorists.'[4]

In at least two of these attacks a .5in-calibre Barrett Model 82 heavy sniping rifle was used; capable of piercing light armour, it has a maximum range in excess of a mile. This had serious implications for the security forces patrolling the border areas: armoured Land-Rovers and soldier's body armour no longer afforded their users protection from such a weapon. British press reports suggested that the IRA sniper(s) might be a renegade from the US Special Forces. These speculations aside, the American connection was underlined when US-manufactured sniper/hunting rifles like the Barret and the Remington Woodmaster were found entering Ireland for use by the IRA. In Crossmaglen, scene of many attacks on the Army, IRA supporters taunted the security forces with a placard reading, 'Welcome to Crossmaglen – Sniper's Alley'.

Snipers in Commonwealth armies had tended to follow the British lead in sniping. In the post-war period, while the basic tactical philosophy remained similar, new approaches to weapon selection were initiated. During the Korean War, Canadian snipers supplemented their No. 4(T) rifles with US M1C and M1D models, and when Canada adopted the FN FAL as its basic infantry rifle, considerable effort was devoted to developing a sniper version. Eventually, however, in 1975, the Canadian Army adopted a completely new sniper rifle – designated the C3 – which was a modified Parker Hale 1200 TX, incorporating a 6-power Kahles telescopic sight. The British-made Parker Hale was a highly accurate rifle designed for sniping, capable of holding five rounds and fitted with butt spacers to allow adjustment for length.

After the Second World War the Australian Army had committed itself to preparing for a war in a jungle or tropical theatre of operations. As a consequence, little formal attention was paid to sniping, although during Australia's involvement in Vietnam soldiers trained with US snipers from the 25th Infantry Division. In the early 1970s, however, Australia's strategic philosophy was modified to incorporate continental defence, which meant greater attention was paid to conventional warfare. Interest was renewed in sniping, and the need for a new sniping rifle to replace the old .303 No. 1 Mk III became obvious. Extensive trials were held in 1978, and the Parker Hale emerged as the most suitable. The new sniper rifle was designated the L3 in Australia, and fitted with a 6-power Kahles scope.

Following British and Commonwealth practice, the South African armed forces were re-equipped with the FN FAL rifle, designated the R1. In the open veldt which is the distinctive landscape of much of southern Africa, the R1 proved an excellent service rifle, providing good long-range accuracy. But the R1 could never replace a proper sniper rifle, and because of the arms embargo imposed on South Africa it was impossible to buy an 'off-the-shelf' weapon. As a result, South Africa turned increasingly towards developing its own weapons, especially in conjunction with Israel. A South African version of the Israeli Galil assault rifle, the R4, was a result of this relationship. A heavy-barrelled 7.62mm sniper version of the rifle was subsequently built, providing the South African Army with a long-range weapon.

On 2 April 1982, Argentina invaded the Falkland Islands. After initial surprise at the attack, Britain's armed forces were swiftly mobilized to form a task force to recapture the islands. The open terrain of the Falklands was dominated by small rocky ridges and hills, an advantage to the Argentinians who were on the defensive, and who also outnumbered the British force. Accordingly, the British planned and executed their attacks as night operations, relying on surprise and superior training to overcome the Argentinian numerical advantage. As a consequence, the vast majority of the fighting took place in darkness, so that sniping became the preserve of those who were equipped with night sights.

The one major exception was the battle of Goose Green, fought by the 2nd Battalion, the Parachute Regiment, against a large Argentinian force holding well-entrenched positions. Although intended as a night attack, the scale of the action ensured that the fighting went on into the next day. The snipers on each side were then able to engage in daylight shooting; the battleground was open and flat and bound on two sides by water.

As the British paratroopers were forced to break cover to advance along the Darwin Isthmus towards Goose Green, they presented relatively easy targets to the Argentinian snipers. A British source noted the accuracy of the Argentinian fire, explaining how a paratrooper 'went back under fire to retrieve [a soldier's] belt, containing 100 rounds. Suddenly, he cried out and fell. He had been shot through the neck by a sniper's bullet, which broke his neck . . . he was dead by the time he hit the ground.'[5]

At the same time, other British troops came under enemy fire. Private Ken Lukowiak's section was pinned down by an Argentinian sniper, one of the paratroopers being hit in the leg. The call went back over the radio to bring up a sniper. A short while later, a ghillie-suited figure arrived at the paratroopers' position. As bullets cracked overhead the British sniper calmly surveyed the Argentinian lines through his binoculars. After locating the enemy sniper's hide, the sniper moved into position and loosed off a single shot. The rifle fire from that part of the Argentinian lines ceased, and after the battle the British sniper found he had shot his enemy straight through the head.[6]

After the British victory at Goose Green the focus of action moved to the Argentinian troops holding Port Stanley. The Argentinians had prepared defences on the mountain tops surrounding the Falklands capital; apart from extensive minefields they had erected stone-built emplacements for their machine-guns and snipers. The Argentinian snipers were despatched to stiffen the forward defences which, with a few exceptions, were manned by poorly trained conscripts. The Argentinian snipers were armed with a mix of weapons, ranging from locally made models of the old German Mausers used in the two world wars to the latest US-manufactured Remington 700s fitted with Redfield telescopic sights.[7]

Argentinian troops were well equipped with night sights, mainly second-generation models such as the AN/PVS–4, which were superior to the first-generation Starlight scopes used by the

British. The infantry on both sides were armed with FN FALs, and the soldiers equipped with night sights on their FNs became the snipers during the final actions that decided the war. The British were impressed with the determination of the Argentinian snipers, who caused many casualties. During the attack on Mount Longdon, an entire British company was held up for hours by a single Argentinian sniper. 'Men found themselves being hit more than once by the same sniper,' wrote one British officer, 'a terrifying tribute to the accuracy of the Argentinian's fire.'[8]

In the fast-moving battles orchestrated by the British, anti-tank weapons (Carl Gustavs or Milans) became an effective, if expensive, counter-sniper weapon. What the rockets lacked in pinpoint accuracy they made up for in destructive power, enabling them to smash the stone sangars that protected the Argentinian marksmen.

A basic tactic adopted by the British was for the soldier with the night scope to fire tracer rounds at the target to guide his comrades' conventional rifle fire. During the Scots Guards' assault on Mount Tumbledown, Lieutenant Anthony Fraser described a successful night sniping incident: 'At one stage, Sergeant Dalgliesh of 15 Platoon shot a man in the head, using a night sight. The Argentinian screamed for at least half a minute – a real caricature, pantomime of a noise. I thought it was a joke, but I found afterwards that he was dying.'[9]

The Falklands acted as a stimulus to sniping in the British Army, confirming its usefulness and encouraging further development. But even before the successful conclusion of the campaign, it was obvious that the old L42A1 rifle – with a lineage dating back to the early years of the century – lacked the refinement and accuracy of the sniper rifles now in the armouries of the major powers. Despite being a leader in sniper training and deployment, Britain had lagged behind in the matter of equipment. Towards the end of 1982 five shortlisted companies submitted designs for British Army trials. One of the conditions of the trials was for a detachable ten-round magazine, a requirement that eliminated most of the contenders. Eventually the choice remained between a Parker Hale (already chosen by Canada and Australia) and a new rifle designed by Olympic shooting champion Malcolm Cooper, and manufactured as the PM by Accuracy International of Portsmouth.[10]

The PM rifle was the winner, and entered Army service as the L96A1. At the time it was regarded by some observers as a controversial choice; the Parker Hale was a first-rate rifle and doubts were voiced over the bolt action of the L96A1. But following its introduction into Army service in 1986, the L96A1 has been generally well received by snipers who have used the rifle. Certainly, it is an accurate weapon: during the trials held at the School of Infantry in Warminster, the L96A1 was able to place ten rounds consistently into a roughly head-sized 38cm by 40cm target at a range of 800 metres.

Accuracy International designed the rifle in four variants. The first is the standard L96A1 infantry weapon, complete with detachable bipod, ten- or twelve-round magazine, iron sights and a 6×42 Schmidt and Bender or telescopic sight. The second is the counter-terrorist variant which has a 12-power scope, flash hider, bipod and a stabilizing leg fitted to the underside of the butt. The third is the 'moderated' variant for use in situations where noise suppression is important, and which comes with a 6-power scope, bipod, moderated barrel and matching subsonic ammunition. The fourth is the long-range variant, which can be chambered for the .300in Winchester Magnum, 7mm Remington Magnum or 8.6mm (.338in) Lapua Magnum rounds. Single-shot only, this variant has been designed for shooting at ranges well in excess of a thousand metres, and a suitably powerful telescope is fitted to the rifle, according to customer preference. These four variants give the PM sniper system great flexibility, and are especially useful for special forces units operating in differing sniper environments.

Since the early 1990s the PM system has begun to be overhauled by the AW series, which consolidates a number of improvements made to PM rifles, notably an improved bolt action and more powerful sights. The L96A1 will, however, remain in service with the British armed forces for some time, and it is a suitably accurate sniper rifle to enter the next century, in keeping with the professional level of training and combat experience of British snipers.

CHAPTER ELEVEN

THE BASICS OF SNIPER TRAINING

Modern sniper training is essentially a consolidation of lessons learned in the past, from the trenches of the First World War to the jungles of Vietnam. But whereas in the past training was often haphazard, today it is systematic in approach and incorporates many of the latest developments in infantry technology.

The theory and practice of sniping exists at the highest levels in the armed forces of Britain and the United States, and consequently their training sets the standard to which others aspire. Since the 1970s, the two nations have developed a close relationship in training and operational matters, and as a consequence there is a broad similarity of approach to sniper instruction. Both Britain and the US share a similar philosophy regarding the mission or role of the sniper, which is expressed in this definition: 'An infantry soldier who is an expert marksman and observer with the ability to locate an enemy, however well concealed, then stalk up or lie in wait to kill him with one round. He is able to observe, interpret and accurately report enemy movement. He can observe without being observed – kill without being killed.'[1]

Before training can begin a thorough selection process is undertaken. Not only does this find the right men for the course, it eliminates unsuitable candidates, thereby reducing wastage. The demands made upon the sniper are sufficient to ensure that only a few above-average soldiers will be capable of meeting the required standard. As a consequence, candidates need to be capable soldiers who have mastered all the skills of the infantryman, especially fieldcraft. The master sniper in the candidate's parent unit can play a useful role in screening men for sniper school. The recommendation of the commanding officer is also important; his co-operation is essential if first-rate men (with at least a year's service remaining) are to be put forward for training.

The first of the specific qualities and skills required of a sniper is expertise in marksmanship, preferably demonstrated by high scores in shooting contests or qualification as an expert marksman. Good physical condition is also necessary, as extended operations require the sniper to go without sleep, food or water for long periods. He needs good hearing and excellent, unaided vision; the wearing of glasses is a liability, as the lenses can reflect sunlight and loss or breakage would render the sniper ineffectual. Smokers are discouraged: smoking on a mission makes the sniper vulnerable to detection, while to refrain can lead to nervousness and irritation, thereby lowering efficiency.

Mentally, the sniper needs intelligence, initiative, common sense and patience. On the training course, he has to master a wide and often disparate body of knowledge, and in combat he must be able to apply this information swiftly and correctly. Lastly, the sniper must be a mature personality who is able to cope with the stress of calculated, deliberate killing. A US Army training circular explains: 'The sniper must not be susceptible to emotions such as anxiety or remorse. Candidates whose motivation toward sniper training rests mainly in the desire for prestige may not be capable of the cold rationality that the sniper's job requires.'[2]

MARKSMANSHIP

Although sniper courses vary in emphasis and structure, they are divided into three broad subject areas: marksmanship, fieldcraft and tactical employment. The first of these concentrates on improving the candidates' shooting skills and focusing them on to the sniping environment.

The high-velocity rifle bullet has enormous power but like all objects in flight it is subject to the forces of air resistance and gravity. Correctly fitted sights account for these forces over

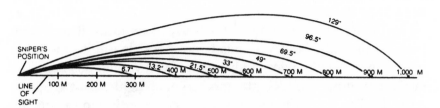

Figure 1: Trajectory of a rifle bullet.

varying distances; the firer adjusts the sights to the appropriate range and, all being well, the bullet describes a curving trajectory that will hit the target. As the bullet leaves the barrel it rises progressively to a maximum height – the culminating point – at about two-thirds distance, before dropping down towards the target.

When 7.62mm rounds are fired at relatively short ranges the bullet has an almost flat trajectory, so that at 300 metres its culminating point is only 6.7in above the line of sight. As ranges increase, the height of the trajectory increases at a geometric rate, and at 1000 metres the culminating height is an extraordinary 129in above the line of sight. The significance of this is crucial for the sniper: because he is normally firing at long ranges he must be able to judge distances with precision. Too short and the bullet will hit the ground in front of the target, too long and the bullet will fly over the target's head.

In order that the sniper can look through his sights and ensure that he will hit what he sees, his rifle must be correctly zeroed. Telescopic sights are fitted with elevation and windage turrets: the elevation turret is adjusted to take into account differing ranges, and the windage turret compensates for lateral (left to right) movement in the flight of the bullet. The zero of a rifle is the elevation and windage setting required to place a shot in the centre of a target – at a given range – when no wind is blowing. By using fixed conditions on the shooting range the sniper can adjust his sights to ensure that at a series of ranges (100-metre intervals, for example) the bullet will strike the target. The zero will slip over time – and if the sights are knocked or

175

otherwise maltreated – so the rifle and sights will periodically need to be re-zeroed.

In theory, a bullet aimed at one spot should hit it time after time. In practice, however, this is not possible due to the interplay of external variables, and the shots will vary in position. The marksman none the less attempts to get as tight a shot group as possible. The size of the group will vary according to the quality of the rifle, the consistency of the ammunition and the skill of the firer, especially in his understanding of such variables as range and wind. A skilled marksman firing match-grade ammunition in a sniper rifle will be able to group his shots within a diameter of 25mm at a range of 100 metres. As range increases the shot group will increase proportionally, so that at 600 metres the group will be 150mm – sufficient to ensure a first round kill.

Once the student is familiar with his rifle and telescopic sight he will attend the range to improve his shooting skills. There are four basic firing positions: standing, kneeling, sitting and prone. Whenever possible the sniper should adopt the prone position, which provides good support and presents a low silhouette to the enemy. Standing, by contrast, affords good vision but leaves the sniper vulnerable to observation, and because of the weight of the rifle, accurate aiming becomes increasingly difficult.

On the battlefield the sniper should adopt whatever position allows adequate observation of the target and maximum support. Aids to support include the bipod – fitted as standard to sniper rifles such as the Parker Hale M85 and the L96A1 – and the sling. By correctly bracing the sling around the forearm, the stability of the rifle is improved, even when firing offhand (when the rifle is unsupported), and the effect of recoil is reduced.

The rifle must be firmly placed into the pocket of the right shoulder (assuming the firer is right-handed), to ensure a steady position and to avoid excessive recoil. The left hand holds the forestock steady in order to keep the sights on target and to prevent canting, movement to left or right. If a bipod is being used then the left hand is brought back to the butt stock, forcing it against the shoulder. The right hand holds the small of the stock firmly but not rigidly, while maintaining rearward pressure against the shoulder. The trigger finger is positioned on the trigger so there is no contact between the finger and the side of the stock. This enables the trigger to be pressed fully to the rear without disturbing the aim. Where possible, the thumb of the right hand extends over the small of the stock to permit the sniper to obtain a spot weld – the point of contact between the sniper's cheek and thumb on the small of the stock. The firm contact between head, hand and rifle enables the head and weapon to recoil as one unit, facilitating rapid recovery after firing. If the sniper is unable to obtain a spot weld he should use a stock weld, where the cheek is placed directly against the stock.

Having taken up a firing position the sniper needs to feel comfortable; he may be there a long time, and as discomfort increases so accuracy declines. A good shooting position demands relaxation from the firer; muscular strain will lead to the rifle trembling, hence the need for adequate support. The body position should be such that the rifle points naturally at the target. To obtain this, the sniper takes aim at the target, then closes his eyes and relaxes. If, when he opens his eyes, the cross hairs of his scope remain on target, then he has a natural point of aim. By moving his body the sniper should be able to adjust his natural point of aim to fit the target.

Once the rifle is being held in a stable and comfortable manner, the sniper must find the correct eye relief, that is the distance (two to three inches) and position between his eye and lens of the scope. Without the right eye relief the shot will not be on target, and if the eye is too close it may be hit by the scope on recoil. Correct eye relief is fairly easy to ascertain, as without it dark shadows will appear around the sight picture. If the shadow is a complete circle, the eye is too far away; if the shadow forms a crescent shape then a bullet fired will hit the target on the side opposite the crescent. By moving his head the sniper can adjust the picture until a clear image is found. Many modern sniper rifles are fitted with an adjustable butt stock, which can be moved to ensure a comfortable position and correct eye relief.

When the sniper has estimated the distance to the target and adjusted the elevating turret and corrected the windage turret, he is ready to begin the firing sequence. Mastering breath control is important if the rifle is to be kept on target; as he breathes so

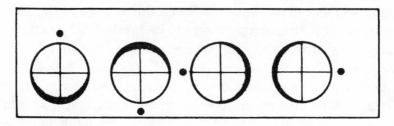

Figure 2: Shadow effects (the bullet strikes on the side opposite to the shadow).

the sniper's chest rises and falls, moving the rifle. Different marksmen have different ways of overcoming this problem – a common method is to fire at the end (or near the end) of the respiratory cycle (i.e. after exhalation) when the muscles are at their most relaxed. The firer's breath can be held for several seconds without affecting the shot but whatever technique is adopted, the emphasis must be upon relaxation. If the aim starts to move around, the sniper should start again.

Trigger control is the next stage in the firing sequence. The slack is taken up in the trigger movement, as the sniper comes to the right point in his breathing cycle and fine-tunes his aim. The trigger is then gently squeezed without moving the rifle in any other way.

The final stage is the follow-through, which prevents the rifle being disturbed while the bullet is still travelling up the barrel. The sniper must keep his cheek in firm contact with the stock; maintain his finger on the trigger all the way back; continue to look through the scope to observe the fall of shot, if possible; and avoid reacting to the recoil and/or noise of the shot. Good follow-through allows the rifle to fire and recoil naturally, to make the sniper/rifle combination react as a single unit to the firing action.

As well as mastering the technical aspects of controlling the rifle, the sniper needs to develop the right attitude. He must focus all his powers of concentration on the shot he is taking, as if his life depended on it. Complementing this concentration is the requirement for detachment from all other distractions. The psychological aspects of marksmanship have been described by the Vietnam sniper and sniper-trainer Carlos Hathcock. Lecturing to a class of sniper candidates about to go on the range on the importance of detachment from their physical surroundings, he made these comments:

> It never rains on the range. It never gets cold or hot. You aren't hungry or thirsty. You are never uncomfortable or lose your attention for any reason when you are in position. When you can mentally remove outside influences from your concentration, you are in your 'bubble'. When I was in Vietnam, the temperature would soar to 120 degrees. It could pour down rain or the wind could blow the stink of rotting vegetation into your face. But I'd get in my 'bubble' and wouldn't notice these things. My only thought was directed towards that one well-aimed shot.[3]

Even when the sniper has mastered the action of firing the shot he must be aware of the influence of external factors, primarily wind and climate. Wind movement is a particular problem for snipers: the effect increases geometrically as range increases. Thus, for example, in a constant 15 m.p.h. wind, a 7.62mm M118 bullet drifts 4.5 inches off-target at 200 yards; at 400 yards this has increased to 20 inches and at 600 yards the figure is 48 inches.[4] Consequently, if the sniper is to hit his target at 'sniper ranges' (beyond the 300–400-yard maximum of the assault rifle) he must be adept in understanding the nature of wind.

The first requirement is to ascertain the strength of the wind. One method is to hold a piece of paper or grass at shoulder level and then release it. The sniper then points his arm directly at the spot where it has landed, and divides the angle between his body and arm by four; the resultant figure is the approximate wind speed in miles per hour. In the field the sniper usually cannot stand up and the wind description method is found more useful. Winds less than 3 m.p.h. can hardly be discerned by the individual, although smoke drifts. Between 3 and 5 m.p.h. wind can just be felt on the face. A 5 to 8 m.p.h. wind will keep leaves in a tree in constant motion. An 8 to 12 m.p.h. wind causes dust to swirl around and blows loose paper. At 12 to 15 m.p.h. wind will be strong enough to cause small trees to sway.

A difficult yet more accurate method of determining wind

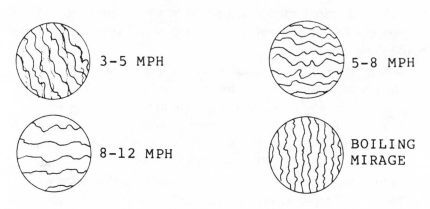

Figure 3: Types of mirage.

speed is achieved by using the spotting scope to observe mirage, the reflection of heat through layers of air at different temperatures and density. The sniper looks through his scope at his target, slightly off focus, and will see a shimmering or wave-like motion. If there is no movement of air the waves will appear to ripple straight up from the ground, a 'boiling mirage'. If there is any wind the waves will be bent in the direction of the air flow (although if the wind is blowing fore and aft it will also show a 'boiling mirage'). Wind speed can be discerned by the wave shape (big waves equals slow; small, flatter waves equals fast) and by the degree to which the wave direction has changed from the vertical to horizontal. By constant use of the spotting scope the sniper will be able to work out wind speeds up to a maximum of about 12 m.p.h.; above that the movement of the mirage becomes too swift to detect for all but the most skilful range shooters.

After determining the wind's speed, the sniper ascertains the effect it will have on his shot. The value of the wind changes according to its direction relative to the sniper's line of fire. A clock system is used to indicate the effect of wind direction. A wind blowing at right-angles to the line of fire is a full-value wind (8,9,10 and 2,3,4 o'clock) while an oblique wind (1,5,7, 11 o'clock) is accorded only half-value, so that, for example, a 10 m.p.h. wind is rated at 5 m.p.h. A wind blowing from the front or rear

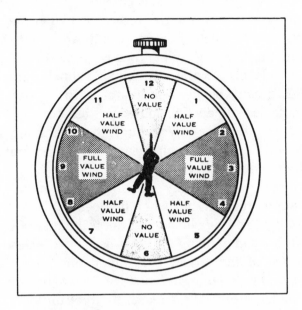

Figure 4: Effects of Wind: The Clock System.

(6, 12 o'clock) is rated as no value, its effect on the bullet only minimal – a head wind slows it down slightly, a tail wind slightly increases its velocity.

When the speed and value of the wind have been determined the sniper adjusts the windage turret, according to a number of set formulas. A common formula – used by the US Marines – is range (in hundreds of yards) multiplied by wind velocity (in m.p.h.) over 15 – a pocket calculator is standard sniper issue. The effects of wind are made more difficult by the fact that air movement is not constant, especially over long distances, and will be stronger in one place than the other. Only experience of shooting in different conditions will provide the sniper with the essential knowledge of how to deal with the consequences of wind drift.

Other effects of weather are not so pronounced but the sniper needs to be aware of their action. A rise in temperature increases muzzle velocity which will make the bullet strike higher on the target. Consequently, with an increase in temperature the sniper

181

lowers the elevation setting, and conversely, for a decrease in temperature he will raise the elevation. The higher the humidity, the denser the air, which offers more resistance to the bullet. To compensate for this the sniper needs to raise the elevation. Different light levels can affect the sniper's aim, but this phenomenon varies greatly between individuals. In considering these weather variables snipers will sometimes tend to overemphasize their influence, and it is only through the careful monitoring of weather conditions that the sniper can derive an accurate appreciation of their genuine effect.

The sniper will normally use his elevation and windage turrets to adjust his aim, but when he does not have the time to change his sight settings he adopts the technique of holding (or aiming) off. Instead of fixing the cross hairs of his telescopic sight on the centre of the target, the sniper aims above or below (elevation), and/or left or right (windage). The sights on a sniper's rifle may be fixed on 500 metres. If a target presents itself at a further distance, the sniper will aim high (at 600 metres, for example, the sniper aims 25 inches above the target centre). If the range is less then the sniper aims low (at 400 metres the sniper aims 15 inches below the target centre).

The sniper is trained to know instinctively the appropriate holds at ranges other than the standard setting. Similar holds can be made to take into account the effects of wind. The mil-dot reticle (cross hairs) in a number of modern sniper rifle telescopes can be used as an aid to holding off, but constant practice is the best method for gaining proficiency. Even so, the sniper rarely achieves pinpoint accuracy when holding off, and this technique should only be used when the time factor is crucial.

Sometimes the sniper has to fire on moving targets, and as a consequence he has to adopt a similar aiming process to that of holding off for wind movement. Indeed, to engage a moving target the sniper must work out the target distance and the wind effects on the round, and also compute the lateral speed of the target and the bullet's flight time (it takes roughly one second for a bullet to travel six hundred yards and roughly two seconds to travel a thousand).

These added variables greatly increase the chances of a miss, and this type of shot is only used when there is no other option. Engaging a moving target at over 500 metres range is not considered realistic. Even at 300 metres a walking target (3 m.p.h.) moving at right angles to the sniper will require a lead of nearly twenty-four inches; for a running man (6 m.p.h.) the figure is doubled. In most instances the target will not be conveniently moving at right angles to the sniper. To allow for this the sniper adopts a rule-of-thumb method to estimate the angle of movement: when the target is moving diagonally to the sniper (nearer 45 than 90 degrees) the lead value is halved; when the target is moving away or towards the sniper (nearer 0 than 45 degrees) no lead is used.

Two aiming methods are employed, according to individual preference, to engage moving targets. Tracking involves the sniper swinging the rifle ahead of the target and firing when the correct lead has been estimated, without any check to the movement. Trapping involves the sniper establishing an aiming point (incorporating the required lead) ahead of the target and firing when the target is the right distance from the point.

Although the sniper has to master all the complexities of long-range shooting individually, he receives help from his spotter/observer. The spotter, looking through a 20-power scope, is better able to observe the target and can correct the fall of shot if the sniper misses with his first round. A key factor here is the necessity of the sniper and spotter developing a good working relationship; poor communication will severely hamper the harmonious and efficient working of the sniper team.

FIELD TRAINING

While the sniper student is improving his skills on the shooting range, he begins the equally difficult study of fieldcraft. Marksmanship allows the sniper to hit the target at long ranges, but it is only through an understanding of fieldcraft that he can find the target, move into the right position to take the shot and, no less important, be able to retire from the firing position to live and shoot another day. In the same way that sniper marksmanship is a development of the basic training taught to every soldier, sniper fieldcraft is a further, advanced stage of the field skills known to all good infantrymen. Field training is taught in separate sections, for reasons of convenience, but in combat all the essential elements form part of an integrated whole.

In order to get to his target, the sniper must be able to read a map in detail, so that he can plan his line of advance and be aware of the time/distance factor of a particular mission. As part of his duties as a FOO (Forward Observation Officer) he will be able to use his map-reading skills to bring down mortar, artillery and air strikes. Complementary to map reading is the interpretation of aerial photographs, which provides additional information, and is especially useful for indicating recent changes on the ground, or when no accurate maps are available.

Camouflage is the sniper's most effective defensive weapon. There are two types of camouflage which the sniper uses to conceal himself and his equipment. Firstly, natural camouflage, the vegetation or other materials that are native to the given area. The sniper should always use some natural camouflage where possible – tufts of coarse grass tucked into a ghillie suit, for example – but he must be careful to replace it as he moves between areas where the vegetation differs, and not to 'overdo' the effect. Secondly, artificial camouflage, as its name suggests, consists of any form of man-made material that aids concealment.

Armies now routinely equip their troops with camouflage-pattern uniforms. Obviously the type of camouflage pattern will depend on the local terrain; standard British DPM, for example, would be inappropriate in the Norwegian mountains in winter or the Arabian deserts, but is a good scheme for temperate climates. The sniper can improve the uniform by breaking up its outline, attaching strips of cloth in a loosely arranged pattern, augmented by bits of natural camouflage.

Although hot and heavy, the ghillie suit is the most effective mode of personal camouflage in temperate regions. Developed by Scottish gamekeepers (ghillies) in the nineteenth century for stalking on the great Highland estates, the ghillie suit renders a stationary man virtually invisible to within a distance of a few feet. British forces adapted the suit for sniping purposes, and through US Marine/Royal Marine exchanges it was 'exported' to America, where it is now standard. Sniper students make their own suits during their 'free' time away from formal instruction. A good suit can take up to fifty hours to put together, but as US Marine sniper Major Jim Land explains, constructing the suit is 'part of the passage into the sniper community'.

A ghillie suit is constructed from one- or two-piece overalls, usually in a camouflage pattern, on to which netting is firmly attached. The front of the suit is reinforced with a heavy cloth such as canvas, and patches are sewn on to the knees and elbows to make prone movement more comfortable. Then literally hundreds of hessian/burlap strips, in several dull colours, are attached (or garnished) to the netting. The front of the suit is only lightly garnished, to aid cross-ground movement, while the strips applied to the back of the suit hang down sufficiently to cover the sides of the sniper when he is in a prone position. Headgear in the form of a bush hat or cowl-like hood is similarly decorated, with a mask or a large face veil attached, that can be used to cover the sniper's face and rifle when he is in a firing position.

A custom in sniper schools is to take visitors out in the field and explain the advantages of ghillie suits, and then for a ghillie-suited sniper to emerge suddenly a few feet away. What was apparently a clump of rough grass is transformed into a bizarre apparition, part scarecrow, part zombie. (The effect is impressive, but the ghillie suit does not actually render a man completely invisible – movement is always a problem – and snipers need to be aware of overconfidence.)

Ghillie suits are necessarily limited to grass and woodlands, and outside these regions other methods of camouflage are necessary. In arctic areas with complete snow cover (including the trees) a full white camouflage suit is worn; in areas with ground snow (but none on the trees) white trousers with green/brown jackets are recommended. In sandy desert regions with little vegetation, camouflage remains a constant problem and while little can be done to break up shape, the careful blending of tan and brown colours helps concealment. In the jungle, camouflage uniforms with contrasting colours are preferred, supplemented by local vegetation; modified ghillie suits are generally considered too bulky. In urban zones blended colours are worn (if possible), various shades of grey being the most effective.

The sniper's face, neck and hands are covered with camouflage sticks or face paint, where available. These come in a variety of colours, applied in a combination to suit local conditions. The parts of the face that form shadows are lightened and those that are prominent are darkened, forming an irregular pattern which helps confuse the otherwise recognizable features of a human face.

Weapons and equipment need to be camouflaged, especially the rifle, with its long, distinctive shape. Many sniper rifle stocks are manufactured in a camouflage texture, but more effective is the application of scrim which breaks up the outline, although care needs to be taken to ensure that the bolt action, trigger and optics are not obscured.

There can be no precise rules for camouflage and concealment – good fieldcraft according to the local circumstances is the best guide – but the sniper should always be aware of those factors that make him vulnerable to enemy observation, and do his utmost to minimize them. A useful guide – the five S's – includes Shape (minimize all distinctive human features); Shine (avoid surface reflection, whether on skin, optics or metal equipment); Shadow (do not cast unexpected shadows, and keep within other shadows); Silhouette (avoid exposure on crests or other skylines); Spacing (maintain irregular intervals with other soldiers' positions, as in the random spacing seen in nature).

The sniper needs to be constantly aware of camouflage, using terrain, vegetation and shadow to remain undetected. During his return to friendly territory the need for camouflage discipline is particularly important, when fatigue and carelessness can easily undermine caution and good planning – with potentially fatal results.

In spite of the best personal camouflage preparation, the sniper is vulnerable when moving. The human eye is always quick to spot movement, especially when it is incongruous, as in, for example, a moving 'bush'. And a further problem for the sniper is that as he is making the best use of available cover, his view of the surrounding area will normally be restricted. In order to make each move as secure as possible, his line of advance should be prepared in detail, prior to action. When on a mission, the sniper should always assume his area of operations is under enemy observation.

Preparation is the key to successful individual movement or stalking. Where possible, the sniper will conduct a detailed survey of the ground he intends to pass over. Otherwise he must examine the maps and aerial photographs he has at his disposal. The sniper should note down (or, preferably, mark on his map or photograph) confirmed or probable enemy positions, likely fire or observation positions he might adopt, and areas of good and poor cover. Then he will prepare a line of advance, divided into a series of short bounds or legs, each leg ending in good cover to allow him to rest, check his position and prepare for the next move.

Stalking involves different forms of movement, whether walking or crawling. The various types of crawl have different names according to service and nation, but they consist of gradations from the high crawl with its relatively fast movement and relatively high visibility to the stomach or sniper crawl with very slow movement (maybe only a hundred yards an hour) and very limited visibility. An expertly conducted sniper crawl is surprisingly effective in exposed terrain, but because the sniper cannot raise his head (it is pressed sideways against the ground), his own visibility is poor and it is easy for him to go astray. The sniper must accordingly be aware of the sun's position, for example, or carefully follow prepared bearings on a simple wristwatch compass. A successful stalk lies in knowing what movement is suitable for the terrain, and matching each leg with an appropriate type of movement.

While planning the mission the sniper would be foolish not to prepare a withdrawal route, preferably separate from the line of advance. The withdrawal plan needs to be as detailed as any other part of the mission, especially if a shot has been fired and enemy troops are alerted to the sniper's presence.

Many of the same principles used during daylight stalks are applicable to night operations. The main differences lie in less need for cover (except the avoidance of silhouettes on skylines) and greater care taken in navigation (especially using a compass). As sound travels further at night, the sniper needs to be ever more silent in his movements, and aware of enemy noises. The sniper can aid his own visibility using the many different types of night sight now available. Conversely, he must take into account the fact that the enemy may be similarly equipped, and so exercise caution in breaking cover.

The location of the fire position, or hide, demands all the sniper's ingenuity and intelligence. He must be able to move in and out of his position without being discovered. The position must give adequate fields of fire and observation of the target area and still provide concealment and some cover (protection) from the enemy. As a general rule the position should be at least three hundred metres from the target area, and ideally a natural or man-made obstacle should lie between sniper and target, so that if the position is discovered it allows him time to withdraw to safety.

An apparently excellent position may also appear that way to the enemy. Thus the sniper has to site his position away from obvious, and especially isolated, locations that will instinctively draw the enemy's attention. It is often best for the sniper to adopt a position in the vicinity of an over-obvious location, so that it will act as a dummy when the shot is fired.

The type of position occupied by the sniper varies enormously according to a variety of factors, such as time, problems of construction and location of the enemy. At its simplest, the position is adopted for a short time (hours, maybe even minutes), sufficient for the sniper to take the shot before withdrawal. In this instance, only some discreet trimming of vegetation is required to secure a better view of the target. It is always important to conceal the muzzle of the rifle, preferably by drawing back into vegetation which will hide and dissipate muzzle flash and blast. Wetting the ground directly below the muzzle or putting down a square of dampened material will prevent dust/debris being raised by the muzzle blast.

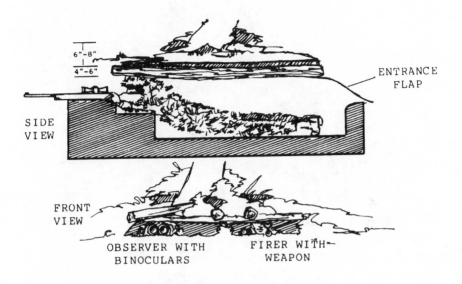

Figure 5: Belly Hide Position.

When the position is to be occupied for a longer period, a more elaborate hide can be constructed – digging a belly scrape, for example – and if time and conditions permit, overhead cover, using local materials, can be assembled. This has the advantage of supplying some protection from enemy fire and improving overall concealment. In addition, it allows the sniper team to move about a little without giving themselves away (remaining motionless for more than a couple of hours can become an agonizing experience, in which mere discomfort turns into genuine pain).

Developing good observational skills is an essential requirement for successful snipers, both to find targets and for general intelligence-collecting duties. Besides an excellent and trained eyesight – a formidable tool in itself – the sniper team is well equipped with optical aids. Typically, they will include binoculars, normally relatively low-powered but with a wide field of view (6×50, for example), which have good light-collection properties and are used for the scanning of potential targets. Snipers often equip themselves with mini-binoculars, as a sup-

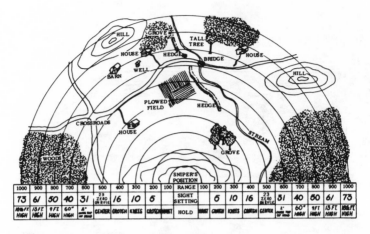

1000	900	800	700	600	500	400	300	200	100	RANGE	100	200	300	400	500	600	700	800	900	1000
73	61	50	40	31	23 ZERO ON RIFLE	16	10	5		SIGHT SETTING	5	10	16	23 ZERO ON RIFLE	31	40	50	61	73	
108 FT HIGH	13 FT HIGH	9 FT HIGH	60" HIGH	2" OVER TOP OF HEAD	CENTER	CROTCH	KNEES	CROTCH	WAIST	HOLD	WAIST	CROTCH	KNEES	CROTCH	CENTER	2" OVER TOP OF HEAD	60" HIGH	9 FT HIGH	13 FT HIGH	108 FT HIGH

Figure 6: Prepared Range Card (drawn from sniper's position).

plement to the standard pair; they are light and small enough to be tucked into a pocket, but have a limited field of view and low brightness.

For detailed examination of specific target areas the 20-power spotting telescope is employed, able to identify objects invisible to the naked eye and barely distinguishable to binoculars. Lastly, the telescope on the rifle (anything from 6- to 10-power) is an effective observational tool. The sniper and observer work as a team to cover and pinpoint potential targets, using each optical device for its appropriate task.

When the sniper team has established itself in its position, it will begin the process of observation. Initially a 'hasty search' of the area is carried out, lasting perhaps thirty seconds. This involves making quick glances at specific points, and is intended to detect enemy movement in areas likely to be a potential danger to the sniper. A 'detailed search' follows, the sniper conducting a systematic examination of a given arc of responsibility. The search begins nearest the sniper in a series of sweeps from left to right, covering about fifty metres in depth. The search extends outwards as far as the sniper can see.

While the search is being conducted the sniper will draw a range card, a rough sketch of the area recording prominent features and distances. The team members should alternate observational tasks to rest their eyes. Patience is the key to successful observation; it may take hours to spot evidence of a well-concealed enemy, but usually some tell-tale sign will be revealed to the conscientious observer.

During night operations the sniper must be conscious of the time needed for the eyes to adapt to darkness (approximately thirty minutes), and the improvement to perception allowed by off-centre vision, where objects are looked at from an angle. The introduction of night-vision aids has greatly extended the duration of activity open to the sniper, although the extension of operations to day *and* night puts him under extra strain – a fact which commanders should not ignore when deploying snipers.

When an enemy target has been found the sniper must decide what action to adopt; rather than take the shot, it may be better to call up indirect fire (mortars, artillery, air strikes) or hold fire altogether and place the target under continuous observation for intelligence-gathering purposes. Experience – especially a well-developed sense of preservation – will guide the sniper to choose the right option. If direct fire is decided upon, the sniper must be prepared for enemy reaction: can he eliminate the enemy threat outright, and if not, can he escape to safety? He must also decide whether to shoot enemy personnel or fire on hardware, such as weapon systems, vehicle optics and communications and radar equipment – all of which, because of their increased sophistication, are increasingly vulnerable to a sniper's armour-piercing bullet.

Key personnel can usually be recognized by actions or mannerisms, positions within formations, rank or insignia and special equipment. Enemy snipers represent the greatest single threat to the sniper, and thus will top his list of priority targets, followed by scouts and dog tracking teams. Officers and NCOs are also priority targets; their loss will slow enemy operations and will lower morale, especially if the troops are poorly trained and need constant direction. Other valuable targets include vehicle commanders and drivers, communications personnel and support-weapon crews. As snipers are the ultimate target-selection weapons, they can afford to leave the rank and file enemy soldier to less discriminating systems, and concentrate on targets that really count.

Once the sniper has located his target he needs to know its range. This is particularly important at long ranges where bullet drop is more pronounced (and, of course, range estimation that much harder), so that at eight hundred metres, for example, an error of fifty metres will probably lead to a missed shot. The most obvious and accurate method of judging target distance is simply to use a map (or air photo), although this presumes the sniper knows precisely his own and his target's position.

Optical devices are another preferred method. Many military pattern binoculars and telescopic sights have a reticle that allow objects of a known height (a 5' 8" man, for example) to be read off against a scale (part of the reticle) and the distance computed. The ranging scales of the Redfield AccuRange scope and the PSO–1 fitted to the Soviet Dragunov SVD are easy to use. The mil-dot system utilized in the scopes fitted to the US M40A1 and M24 rifles is particularly accurate, although more difficult to compute. Other devices include commercially manufactured range finders which work on the principle found in single-lens reflex cameras: once the image comes into focus the range can be read off from a dial. The most accurate way to judge distance is by using a laser range finder, but they are too heavy and too expensive for normal sniper use.

Outside optical/electronic means, the human eye can be trained to estimate range accurately, and sniper instructors work hard to develop this ability in their students. Various techniques are employed to aid estimation. The 100-yard method is good to a range of about 500 yards, whereby the sniper estimates the range in 100-yard units (based on the familiar length of a football field or running track straight). Beyond 500 yards the sniper should choose a point halfway to the target and determine the number of 100-yard increments to that halfway point, and then double it to find the overall range. The appearance-of-objects technique works well in conjunction with the 100-yard method. Range is determined by the target's size relative to other nearby objects. This system presumes the sniper is familiar with the characteristic details of these objects as they appear at different ranges. In the bracketing method, the sniper assumes that the target is no more than an 'x' distance, and no less than 'y' distance, so that an average of both distances will be the appropriate estimated range.

Various external factors influence the sniper's perception of distance, and they must be taken into account before arriving at a final figure. They can be divided into three broad groups. First, the nature of the object: an object with a regular outline, such as a building, will appear closer than one with an irregular outline, such as a bush; an object that is in strong contrast to its background will appear closer than it is; a partially exposed object will appear further away than it is. Second, the nature of the terrain: looking uphill makes the target appear closer, while looking downhill makes it look more distant; as the sniper's eye follows the contours of the terrain he will tend to overestimate distant targets, while observation over flat, featureless terrain, such as snow or sand, will lead to an underestimate of distant targets. Third, light conditions: the clearer the light on the target, the closer it appears; when the sun is behind the sniper the target seems nearer but when the sun is behind the target it is more difficult to see and thus appears further away.

As the sniper team will be operating on the forward edge of the battle line, snipers are trained in the basics of intelligence gathering and reporting. Once in the field they must always be on the alert for the collection of information. A formal observation log of any enemy activity will normally be kept as a matter of routine, supplemented where necessary with military field sketches. Although a modicum of artistic ability is useful, almost anyone can be taught the basics of sketching, and this discipline is taught to most snipers as a useful visual adjunct to an observation log. Increasingly, small cameras are carried by one member of the sniper team, and although a photograph provides a precise picture it has not replaced the field sketch, which is better able to bring out the salient features of an object.

A final aspect of field training is survival, evasion and escape. Because the sniper operates forward of his own lines, he needs a thorough knowledge of survival skills: how to stay in good condition in a variety of hostile environments, and how to deal with problems in the most effective manner. As a sniper may well be cut off from the support of friendly forces, he will need to develop all the skills of evasion, allowing him to move through enemy territory without being caught. And if captured, he must know how best to escape. A central component of survival, evasion and escape training is how to avoid panic, keep calm and develop a simple plan of operations which will maximize the sniper's own resources.

SNIPER EMPLOYMENT

One of the more difficult areas of sniper training lies in the correct employment of snipers on the battlefield. This involves not only the snipers but the platoon, company and battalion commanders who oversee their operations. In the past, infantry commanders seldom had much knowledge or interest in sniper operations, and often regarded them as a distraction from the main business of the infantry battle.

The situation has improved, but sniper employment – where and how the sniper fits in with other troops and what his role should be – remains a difficult tactical problem, requiring common sense and imagination from both snipers and the infantry officers employing them. They need to work in partnership if they are to operate effectively. A recent US Army manual makes the point that infantry officers, '. . . must remember that a sniper mission cannot be tied to a rigid time schedule, nor can a sniper team be effective in positions that will receive fire from the enemy. The team must be positioned far enough away to avoid this fire, but close enough to deliver precision fire at the most threatening targets.'[5]

Although by the nature of their work snipers should operate within their own tactical framework, they none the less do so alongside, and in support of, other troops. Thus, despite the need for freedom of action, they must ultimately conform to the current tactical situation.

A sniper and observer of the 1/4th Royal Berkshire Regiment survey enemy territory from the roof of Anton's Farm near Ploegsteert Wood, spring 1915. (*Imperial War Museum*)

A scene from a trench at Gallipoli held by Royal Naval Division and Anzac troops, April–May 1915. While the officer observes through a periscope (left) another soldier sets up a sniperscope at the forward section of the trench. (*Imperial War Museum*)

Spot the sniper! A demonstration at the First Army Sniper School of the advantages of good camouflage on the Western Front, 1917–18. The soldier on the left, wearing a service cap, is clearly visible, unlike the man in camouflage robes on the right – the muzzle of his rifle can just be seen. (*Imperial War Museum*)

A sniper from the Durham Light Infantry uses a scout telescope to observe enemy movement in the Tilly–Caen area, Normandy, June–July 1944. (*Imperial War Museum*)

Private Francis Miller, a sniper in the 5th East Yorks, takes aim for the benefit of the camera, northwest Europe, 1944–45. At the time the photograph was taken Miller had seventeen confirmed kills and another six 'possibles'. (*Imperial War Museum*)

US Marine snipers in action on the island of Okinawa, 1945. They are firing .30–06 Springfield 1903A1 rifles, equipped with 8-power Unertl telescopic sights. (*National Archive*)

Armed with a .303in No. 4 Mk I (T) rifle, a British sniper prepares to fire from a sangar in Aden, 1964. The wooden cheek rest, attached to the comb of the stock to provide correct eye relief, can be seen in this photograph. (*Imperial War Museum*)

This US Army sergeant in Vietnam holds an accurized 7.62mm M14 sniper rifle with a Redfield variable-power telescopic sight. (*National Archive*)

A US Marine sniper in Vietnam adopts one of the classic shooting positions while scanning the distance for enemy snipers. He is armed with a .30–06 Winchester Model 70 with Unertl scope (*National Archive*)

A view of a US Marine sniper team in action, during Operation Swift in Vietnam. While the observer looks at the target with his spotting scope, the sniper takes aim with a 7.62mm M40 rifle with Redfield scope. (*National Archive*)

Kitted out in Ghillie suits and armed with a 7.62mm L96A1 rifle, a British sniper team examines a potential target. (*Military Picture Library*)

In an offensive action, the sniper acts to eliminate targets that might slow the advance. Operating on the flanks of the attack, sniper teams can deliver precision fire in support of the advancing troops. The enemies' crew-served weapons, including mortars, machine-guns and rocket launchers, are a first priority; knocking them out removes a major element of the enemy's firepower at the critical moment when friendly troops have to move from cover towards the enemy. Also, the personnel of enemy crew-served weapons often have to adopt vulnerable positions. The operator of a wire-guided anti-armour missile, for example, is dangerously exposed while he guides the missile to his target – sufficient time to be hit by a sniper's bullet, or failing that, to lose his concentration and miss the target.

Snipers can be used effectively in a number of quite specific tactical situations. In river crossings, a sniper team can occupy a commanding position on the river bluff or other high ground some distance from the river, and yet still deliver accurate fire to the far bank. Tanks and other AFVs (Armoured Fighting Vehicles) are often vulnerable to ambush by determined snipers. Unless there is an obvious and visible threat, few AFV crews will operate with their hatches battened down. This gives the sniper the opportunity to kill the commander or driver, whose heads are visible outside the AFV; one well-placed shot can bring a fifty-ton tank to a dead halt. Even when shut down, the AFV's optics remain vulnerable, so that if the driver's vision block has been shattered the vehicle is severely impaired. Similarly, many essential electronic sensing devices (infra-red, radar) are exposed, and are easily damaged by an accurate shot from a knowledgeable sniper.

In defensive operations, the sniper's role is to shoot specific targets to delay the enemy's advance. The elimination of officers and NCOs becomes particularly important, especially against poor-grade troops; without direction they invariably halt and, in the face of accurate fire, will go to ground.

Defensive fighting is often a forced and hasty business, and the sniper must fit in as best he can. When, however, there is more time to establish an organized defence, sniper teams will work alongside each other to provide mutual support, and so be correspondingly more effective. If a withdrawal or retreat is forced upon an army, then snipers are especially valuable in holding off the enemy. Their infantry will not press home the attack with much alacrity if they know there are well-trained snipers in the vicinity. Also, knowledge of a sniper's presence will force the enemy to deploy into a slow-moving combat formations, further impeding progress.

Fighting in built-up areas offers good concealment and cover to both attackers and defenders, and poses new challenges to the sniper. Observation and fields of fire are clearly defined by streets and buildings, but surveillance is made more complex by the numbers of observation and fire positions on rooftops and in windows and doorways. While observation of the enemy is made more difficult, similarly the sniper is difficult to locate. Dummy positions are frequently used to draw enemy fire away from the sniper's position. The short ranges over which most combat in towns and cities is fought tends to negate the sniper's range advantage. The sniper can overcome this, however, by the application of his superior knowledge of fieldcraft and, where possible, by firing from positions to the rear of the infantry combat zone but within range of his own rifle.

Movement in built-up areas needs particular care. Activity at street level in the open is easily detected, so the sniper should, if possible, smash his way through building interiors – mouse-holing – or use underground passages, such as sewers or water-ways. When taking up fire positions, the sniper needs to be aware of the special circumstances of this artificial terrain. When firing out of a window, he should stand well back in the room, muffling the blast and hiding the muzzle flash. Windows and doorways are obvious firing positions; it may be better if the sniper cuts a funnel-shaped hole through the wall. A hole of this nature is hard for the enemy to locate and hit, while giving the sniper a reasonable field of fire. At all times the sniper must keep in touch with the overall tactical situation, so that, for example, he is able to retire to new positions if there is any chance that he might be overrun.

As the greatest danger to a sniper is another sniper, counter-sniper operations are taught at length. It is standard practice for sniper training to be conducted as if there is an enemy sniper present, with the instructors acting as enemy snipers. When a student can out-track and out-shoot a sniper, he can consider himself well versed in the craft of sniping.

The first step in counter-sniping is to ascertain if an enemy sniper is operating in the vicinity. Clues to a hostile sniper's presence include visual identification of a well-camouflaged soldier carrying a bolt-action rifle with telescopic sights, or a sudden increase of single-shot casualties among key personnel. Once his presence has been established, the sniper will gather all available information and attempt to determine any patterns in the enemy sniper's behaviour before drawing up his own plan of action. The sniper will go out in the field, using every scrap of knowledge and cunning to outsmart his enemy, then locate his position and eliminate him.

A thorough training in the three main subject areas – marksmanship, fieldcraft and sniper employment – provides the sniper with a framework on which to develop this complex military skill. Yet, in the end, only the final acid test of war will transform the sniper student into the combat sniper. Instructors must therefore prepare their students in as effective a way as possible for the real thing.

CHAPTER TWELVE

THE SNIPER TRAINING COURSE

The Royal Marines were the pioneers of sniper development after the Second World War, and their training programme has subsequently become a standard among armed forces in the West. The Royal Marines train snipers from around the world; apart from regular US–British exchanges, they instruct British special forces (including the SAS), and troops from NATO countries including Holland and Norway.

The Royal Marine training programme has set a pattern which other services have followed and adapted for their own purposes. It is therefore worth looking at in some detail. The sniper course consists of a highly intensive five or six weeks of instruction, followed by the Sniper Badge Test, an evaluation of the sniper's abilities. The sniper must pass all the test subjects to qualify as a sniper marksman (the top grade) or sniper 1st class. The test is retaken annually by each sniper, to ensure standards are maintained.[1]

The first subject is shooting, and consists of a series of tests of marksmanship, firing at varying ranges from differing positions, at both stationary and moving targets. Because of the shortage of 1000-metre ranges in Britain, maximum range shooting is usually confined to 600 metres for this exercise, although the snipers are trained to fire out to 1000 metres.

The second test is camouflage and concealment (popularly known as cam and con), designed to examine the sniper's ability to locate and fire from a properly camouflaged and concealed position. The cammed-up sniper, wearing a ghillie suit, is given five minutes to conceal himself in a specified area, between 150 and 250 metres from an observer (one of the instructors) equipped with binoculars. Another instructor (known as a 'walker') has a radio link to the observer and when the sniper is concealed he will walk over to the sniper's general position. When the walker is within ten metres of the sniper he will radio this fact to the observer.

All the while, the observer will be looking for the sniper, but if he remains undetected the walker tells the sniper to fire a (blank) shot at the observer. If the sniper still has not been seen, the walker indicates his position by touching his head. Providing the sniper remains hidden from the observer's view, the walker instructs the observer to make a movement which the sniper must identify (for example, a card displaying a number). If the sniper is able to read the number, his sights are checked for an accurate setting of elevation and windage and, if correct, he receives a pass. For the Badge Test, three such exercises are carried out, and the sniper must pass two out of three.

The aim of the third test, stalking, is to demonstrate the sniper's ability to use ground, by approaching and occupying a fire position without being observed. Many of the same skills of camouflage and concealment are applicable. The sniper is required to stalk 800–1000 metres towards an unseen objective within a given time limit (normally two hours). The sniper is provided with a map and an aerial photograph, on which are marked the boundaries and the objective – a notional OP occupied by two observers equipped with binoculars.

The sniper stalks to a fire position of his choice, between 150 and 250 metres from the OP, and then fires a blank. The observers are assisted by walkers who confirm sightings, and who check the suitability of the sniper's fire position. The same cam and con procedures are also used here. The walkers indicate the sniper's position to the observers and, if still unseen, the sniper fires another shot. If he remains undetected, and providing his sight settings and fire position are satisfactory, he passes the test. The sniper fails if he is seen during the stalk or his fire position is observed.

The fourth test, observation, is a method of judging the sniper's ability to observe and record accurately the results of his observations. He is equipped with 6-power binoculars and a 20-power spotting telescope. Twelve articles of military equipment, such as a pistol magazine, the tail fin of a mortar round, a rifle cleaning brush, are positioned in an arc to the sniper's front, at a distance of a hundred to three hundred metres. They are placed so that they are invisible to the naked eye, are just visible using binoculars, and recognizable with the telescope.

The sniper is allowed forty minutes to locate the articles, identify them and accurately plot their position on a panoramic military sketch. Although a seemingly straightforward task, this test is possibly the hardest of all; not only does the sniper need to possess very good eyesight, he must also have excellent powers of interpretation, to realize what the partially revealed articles actually are.

Test number five, judging distance, measures the sniper's ability to estimate range accurately. He must calculate the range (to within 15 per cent of the exact figure) of eight objects (none of them a figure or range target) at ranges of a hundred to a thousand metres. The sniper can use his binoculars and rifle scope to assist his judgement.

The purpose of the sixth test, map reading and aerial photograph interpretation, is to assess the sniper's navigational skills. He is taken to an unfamiliar position, provided with a map reference of that position and instructed to travel to another map reference about two miles away. When the sniper arrives at the position, he is asked to make comparisons of ground features with those on his map and the aerial photograph. He will be expected to give exact bearings and ranges to a number of given points.

The seventh and final test, sniper knowledge, is a written exam covering all aspects of the course. Subjects include the workings of the rifle and telescopic sight, as well as the other optical instruments; ammunition and ballistics; and the scaling and plotting of aerial photographs. The sniper will also be tested on the use of radio and voice procedures; the location, identification and indication of targets for the purposes of bringing down indirect fire and air strikes; NBC (Nuclear, Biological, Chemical) conditions as they relate to the sniper; intelligence, including the organization of friendly and hostile forces, equipment and minor tactics; AFV recognition; combat survival, escape and evasion; and resistance to interrogation.

The Royal Marine instructors based at Lympstone in Devon aim to produce snipers who have a thorough grounding in the disciplines of their subject. The emphasis is placed on the man, rather than technology. This is partly due to a shortage of money for indulging in the latest hi-tech equipment, but it is also a matter of choice: during wartime it is easier to buy in equipment than find and train top-quality snipers. Standards are high but so is demand for the coveted sniper's badge, and there is little difficulty in attracting men to the course.

The British Army sent members from the Small Arms School Corps (SASC) to train with the Royal Marine snipers at Lympstone in 1970.[2] Since then the Army has developed its own sniper courses at the School of Infantry, Warminster. The approach and methods used by the two services are very similar, and there is a close liaison between Royal Marines and Army. One difference, however, lies in the type of sniper student in each service. The Royal Marines directly train privates and junior NCOs to be snipers. The Army trains its soldiers (senior NCOs and junior officers) to be sniper instructors; because of the Army's greater size the Sniper Division at the SASC does not have the 'luxury' of being able to train every sniper directly. Consequently, once the Army snipers have completed the course they will go back to their units to train snipers in the battalion sniper section.

The US Marines have become leaders in sniper training, devoting an eight-week course to scout-sniper instruction at the main training school at Quantico, Virginia (there are two divisional training schools at Camp Lejeune and Camp Pendleton). The Marine Corps Scout Sniper School was set up at Quantico in 1977, under the guidance of Majors Jim Land and Dick Culver, with Carlos Hathcock as the first NCOIC (Non Commissioned Officer In Charge). The formation of the school signalled a radical change in the Marine Corps's attitude to sniping, as for the first time sniping was accorded full peace-time status. No longer would sniper courses be hurriedly set up as a wartime expedient.

Manuals were prepared to form the basis of a Marine sniper doctrine, and Marine snipers were given their own MOS (Military Occupational Speciality) of 8541 Scout Sniper. In the infantry battalions, a scout-sniper section was added to the T/O (Table of Organization).

Since its foundation, the Marine Scout Sniper School has developed into a first-rate sniper training foundation. Not only do Marines pass through the School, they receive students from the Army, Navy SEALs, and even a few agents from the FBI. The training developed by the US Marines is broadly similar in content to that taught to British snipers. In a breakdown of 205 hours of formal instruction, over 62 hours are given over to marksmanship, 53 hours to field training, and 85 hours to sniper employment – the last including extensive field exercises to put all the lessons learned previously into a realistic tactical frame-work. A distinctive feature of Marine training is the attention paid to long-range marksmanship. The Marines possess a fine rifle in the M40A1, and with adequate long-distance shooting ranges they regard 1000-metre shooting as commonplace.

The instructors push their men hard, and a degree of mental and physical stress is deliberately built into the course, with the students facing fourteen-hour days in the field and in the class-room. A former officer in charge of the school, Captain Tim Hunter, explains the reason why: 'This is not a boot camp and we don't treat our students like recruits. They're already good Marines or they wouldn't be here. But we make the course stressful. Being a scout sniper is an extremely stressful business. The school shows us how a future scout sniper will act and react under pressure. If he can't handle that stress in a school situation, there is no way he will be able to cope with it on the battlefield.'[3]

As a result of the tough approach demanded by the School, the drop-out rate is fairly high, as another former officer in charge, Captain Steven L. Walsh, notes: 'We have a relatively significant attrition rate, from people who are not psychologic-ally, emotionally or physically prepared to do it. Eventually they come in and say "I can't do it". We don't drop 'em. They leave.' Walsh emphasizes the need for a stable and cool temperament for a successful sniper: 'We have the capability to watch people die; his head explode or whatever. It's the mark of a true professional to carry out the mission.'[4]

Alongside formal sniper training the School provides what it calls 'confidence builders'. They include a combat pistol course, a Quick Kill refresher course for the M16, knife-fighting instruction and a hand-to-hand combat course based around taekwondo, judo and aikido.

A relatively new development in Marine training has been the introduction of a scout sniper instructor course, which is taught alongside the standard course. The intention is to admit students who are already trained scout snipers (8541s) or STA (Surveillance and Target Acquisition) platoon sergeants and teach them to be sniper instructors, tactical advisers, and 'subject matter experts' on scout sniping within their infantry battalion. As part of a tightly argued article on the challenges facing Marine sniper training in the 1990s, Captain Edward D. Daniel explained the battalion sniper instructor's role: '[He] will train his scouts and scout snipers both before and after they have attended a scout sniper course. The aim of this approach is to ensure that STA Marines are properly prepared for scout sniper schools (to minimize attrition) and to ensure that school-trained Marines stay combat ready.'[5]

The US Army was slow to follow the lead taken by the US Marines in developing peacetime sniping. But, taking their lead from the Marines, a sniper training school was set up at Fort Bragg, North Carolina, during the 1980s, under the auspices of the AMTU (Army Marksmanship Training Unit) of the XVIII Airborne Corps.[6] This tough five-week course – rated as harder than Ranger School – swiftly weeds out those who lack the special qualities of self-reliance that mark out the true sniper. Alongside long-range 1000-metre shooting, the course emphasizes operations in the field, especially in those situations where airborne troops are deployed, far-flung corners of the world with limited logistical back-up.

In 1987 the Army finally organized its own general sniping course at Fort Benning, Georgia. Three weeks long, it is necessarily more limited in scope than that taught by the Marines, but it none the less trains its students to engage targets with precision fire at ranges out to a thousand metres and to be proficient in fieldcraft. The school is run by the 29th Infantry Regiment, and training is demanding, with every moment of the student's time taken up with work. Captain Mark L. Rozycki, who was responsible for the operation and administration of the school in the late 1980s, made these notes on the running of the course:

> Training during the first week is oriented on fieldcraft techniques, sniper patrol orders, and sniper movement techniques. The students also zero their weapons and receive training on marksmanship fundamentals. They spend an average of four hours a night constructing their ghillie suits, writing their first patrol order, and studying. Cadre members are available to help them.
>
> Throughout the second week, the soldiers participate in a number of evaluated exercises that include their first record fire, concealed movement, and target designation. To facilitate training, the sniper class is divided into two training groups but with the members of each sniper team kept together. On the eighth day, the training moves out of the classroom to the range and the fieldcraft training sites. The focus of the training is now placed on advanced sniper marksmanship as well as sniper fieldcraft.
>
> The training continues through the third and final week, which concludes with an arduous 24-hour sniper team exercise. During the tactical exercise, a sniper team is evaluated in a number of areas including sniper patrol order, construction of positions, target reduction, and final shot.[7]

While the various sniper courses cover the tactical conditions a sniper is likely to encounter, members of special forces units will face additional situations where the use of a sniper could be highly profitable. For those special forces soldiers nominated as snipers it is customary for them to receive basic training at a sniper school which best suits their requirements (the SAS at the Royal Marine school, for instance). They will subsequently develop these skills at their home training centres. Some units do, however, develop complete sniper courses of their own. The content of a special forces sniper programme depends on the nature of the unit, and where and how it expects to operate.

Britain's SAS operates in a wider field of action than most special forces units, whether in a counter-terrorist role on city streets, taking part in covert paradrop insertions or performing long-range reconnaissance missions deep behind enemy lines. As a consequence they will have to expand their training to fit the multiplicity of missions. Thus, for example, an SAS sniper could easily find himself providing precision fire support during an aircraft hijack, where other weapon systems could endanger innocent lives. In this context he will familiarize himself with the layout of the standard passenger aircraft, especially knowledge of exits, and have an operational plan at the ready should an incident occur. By contrast, other units have a more specialized role. The Royal Marines' Commachio Group has a prime responsibility for protecting Britain's offshore oil and gas wells, and so will focus training for operations involving sniping in and around oil rigs in the North Sea.

The US Navy's SEAL (SEa Air Land) units pay particular attention to sniping, as a preliminary note to their sniping syllabus points out:

> The future combat operations that would most likely involve Naval Special Warfare would be low intensity type combat operations that would employ SEAL personnel in the gathering of information for future military operations, or the surgical removal of military targets and personnel with a minimal assault force, with no loss of life to civilian personnel. This is the ideal mission profile to employ snipers, due to their advanced field skills, marksmanship and their ability to operate independently in a field environment.[8]

As a result of their commitment to employing snipers in a diverse number of roles, SEALs have felt it necessary to develop a nine-week sniping course. The basics of the course are similar to those of the US/Royal Marines, but there are additions to suit the special nature of SEAL operations. One of the most obvious is the forty hours devoted to naval gun fire support, the SEAL sniper team working ashore and providing FOO (Forward Observation Officer) direction to the sea-borne 'artillery'. The snipers are also trained to support 'across beach operations' and the boarding of potentially hostile vessels at sea. As some of their missions involve taking on small vessels, such as enemy patrol boats, at long ranges, emphasis is placed on mastering .50-calibre long-range shooting.

Specialist instruction is given in sniper operations involving parachute drops and helicopter insertions and extractions. SEAL training in calling down air strikes includes directing rotary and fixed-wing gunships as well as the more conventional strike aircraft. And for troops often forced to work behind enemy lines, extra attention is paid to evasion tactics.

British (and Commonwealth) and US forces subscribe to the basic view that the sniper is an élite soldier who operates in support of infantry operations but who is not directly part of an infantry unit. This approach to sniping is not, however, universal. Other nations, notably the Soviet Union, France and Israel, adopt a strategy in which snipers are distributed within the infantry platoons and squads/sections. The principle here is to have a soldier who can extend the normal range of the infantryman, and his mode of operation is closer to that of a sharpshooter than a true sniper.

While sniper training in these armies is often of a high standard, this manner of sniper employment does not allow the sniper full rein to operate in the most effective manner. The US Army realized the shortcomings of this tactic in Vietnam. When their snipers operated at a low organizational level, results were poor, but when they were given the freedom to act in a semi-independent capacity they became a useful infantry asset. Today, US Army training reflects this decision, and their snipers are all the more effective for this policy.

Despite different approaches to training, it should be recognized that only good soldiers make good snipers. Selection is of the greatest importance in weeding out unsuitable candidates, and the training must be sufficiently rigorous to ensure that standards are maintained and unsuitable men returned to their units. By these means snipers become a genuine élite force – whatever their cap badge – inspiring confidence in the commanders who employ them and in the infantry they support.

CHAPTER THIRTEEN

SNIPING IN RECENT CONFLICTS

In the many wars that have taken place in the world over the last two decades, the crack of the sniper's bullet has been heard to deadly effect: in the cities of Beirut and Sarajevo, in the jungles of Central America and in the deserts of the Arabian Gulf.

The Israeli Defence Force (IDF) had traditionally accorded sniping a low priority. In the Israeli philosophy of warfare – dominated by armour, engaged in fast-moving battles of manoeuvre – there was little place for the sniper. This attitude began to change following Israel's invasion of the Lebanon in 1982, when IDF troops were engaged in protracted ground fighting for the first time.

In the Israeli Army snipers are dispersed within the smaller infantry sub-units, at squad and platoon level, their function to give the ordinary infantry (armed with assault rifles) longer-range firepower. In contrast to the practice in most Western armies, the IDF has chosen not to arm its snipers with a bolt-action rifle but has converted its Galil semi-automatic assault rifle (based on the Soviet AK–47) as a sniper weapon.[1] Before the introduction of the Galil sniper rifle in the mid-1980s, Israeli snipers had to make do with US-surplus accurized M14s and a variety of commercially manufactured models.

The Galil sniper rifle has been rechambered for the 7.62mm NATO round (as opposed to the 5.56mm of the assault rifle version) and given a more rigid forged receiver and heavy barrel, along with a 25-round magazine, bipod and a 6×40 Nimrod telescopic sight. The rifle's most distinctive feature is its folding stock, retained from the assault version. The Galil does not have the accuracy of its bolt-action rivals but is able to fire rapid succeeding shots – considered by the Israelis to be of greater importance than in America and Britain. A disadvantage of the Galil is its weight, which when fully loaded and with scope amounts to 18.3lb.

After a rapid drive northwards into Lebanon in June 1982, the Israelis halted outside Beirut and commenced an air and artillery bombardment of the city. The war degenerated into a stand-off between the Israelis and their PLO/Arab opponents. In the resulting shooting match, the Israelis sent their snipers into the battle zone, where they immediately inflicted casualties among the Arabs, unused to infantrymen able to fire accurately over the three to four hundred metre maximum of the assault rifle. An American trainer of Israeli snipers, Chuck Kramer, gave this account of an action against the PLO in Beirut:

> Most of the good hits were at extreme long range and most of the kills were made at 600–800 meters. These guys [the PLO] were getting very wary by then. I got two on a motor scooter by sheer luck. I was watching this big wide avenue and sure as hell, I saw this big motor scooter coming towards me about a kilometer away with two guys on it. They're both carrying SKS carbines slung over their shoulders, two shopping baskets full of food and these canvas carriers for the RPGs on the front of the motor scooter. I lined them up and . . . [fired] . . . by sheer instinct. It was a classic shot. The round must have gone through the driver and then into the passenger.[2]

The PLO and other Arab militia contesting the streets of Beirut seemed to have no clear idea of the disciplines of sniping. Operating alone, Arab snipers would go out to the front line and blaze away at any Israeli troops they could find. Predictably Israeli casualties were not heavy, although when under sniper fire (even if inaccurate) Israeli soldiers found it an unnerving experience, and attacks would be stalled for hours as attempts were made to destroy a suspected sniper's position with air or artillery strikes.

While the PLO and Israelis engaged in their confrontation in the ruins of Beirut, the United Nations attempted to broker a ceasefire. A multinational force sent to the Lebanon in a peace-keeping role included elements of the US Marines. The peace keepers faced an impossible task in the convoluted political morass into which the Lebanon had disintegrated. And yet, despite the many restrictions preventing UN troops from replying to hostile fire, US Marine snipers were eventually able to operate with some success against the Arab gunmen roaming the streets of Beirut.

Much had been learned since the war in Vietnam, and now that the Marine Corps had a peacetime sniper training pro-gramme in place, its snipers went into battle prepared for action. Post-Vietnam, the Marines had upgraded their sniper rifle. While the M40 (the military version of the Remington 700–40) was an accurate weapon, it remained an adaptation of a sporting rifle and as a result had several combat limitations, in particular, the barrel was too light and the rifle lacked overall 'soldier-proofing'.

During the late 1970s the Marine snipers at Quantico began to look for a replacement rifle. Consideration was given to the US Army's semi-automatic M21 system (an uprated M14), but tests conducted in 1973 to compare the M40 against the M21 were critical of the Army rifle for its lack of accuracy and increasing unreliability. Having surveyed a range of sporting and military rifles, bolt action and semi-automatic, the Marines decided to develop their own bolt-action rifle based on the tried and tested Remington 700 action.[3]

The new Marine Corps sniper rifle was designated the M40A1. A heavy 24-inch stainless steel barrel was mated to a Remington receiver, which was bedded into a fibre-glass stock, previously impregnated with camouflage colours. The advantage of a fibre-glass stock was that, unlike one made of wood, it retained its shape regardless of external weather conditions. During the 1980s the Redfield 3–9 adjustable-power telescopic sight was replaced by a new 10-power Unertl scope. Although the adjustable scope provided a flexible sight picture at different ranges, it was less reliable than a fixed-power scope.

Armed with M40A1 rifles, the Marine snipers in Beirut would have a chance to test their new weapons in combat. A site assigned to Marine control was the Lebanese University, and a squad of snipers from the STA (Surveillance and Target Acquisition) platoon formed part of the security force. Rocket-proof defences were constructed with sandbags on the roof of the building, allowing the snipers a good field of fire over the immediate area. A patch of rough ground lay between the Marine position and a built-up area held by Shi'ite Amal militia. The Marines were under constant, if not particularly accurate, fire from the militia but were unable to reply without specific orders from headquarters. Eventually the order came through, allowing the Marines to return fire. Lance Corporal Tom Rutter was one of the Marines on duty when the University came under Amal fire. The alarm was sounded:

Within seconds, the other snipers had scrambled upstairs, taken their places and begun pinpointing targets. They already had several positions marked on the map and knew the ranges. Rutter and Corporal Crumley scoped the distant escarpments and buildings and soon located where some of the fire was coming from. A lone sniper was behind a berm 600 meters away. As they watched, he poked his head up and fired his weapon at the Marine bunker. That done, he ducked down and changed positions along the berm. He poked up from a new spot and fired another burst.

'I'll keep my eye on this one behind the berm until I can get a shot,' said Crumley, looking through his scope. 'See if anyone else is shooting at us.'

Rutter scoped the area. It was mid-August and he sweated profusely. Salt stung his eyes. He blinked it away, concentrating on every detail as he swept his crosshairs slowly across the shattered terrain. Suddenly something caught his attention. In the firing revetment of a sandbagged bunker on the street, a second set of muzzle flashes erupted. Rutter settled the reticle on the target and fired. His single bullet streaked through the window with pinpoint accuracy. The firing from that position ceased.

Crumley centred his scope on the man behind the berm. He had watched carefully and noted where his target presented himself on each occasion. Now it was payback time. Crumley was not the only sniper who wanted a piece of him.

'I want him. He's mine,' Crumley announced, as he read the range from the 'mil-dot' scale and set his scope for 600 meters. When the man stuck up his head again, Crumley fired. A puff of dirt kicked up in front of the man's face and his head exploded. The Shi'ite flipped over backwards and lay still.[4]

As a result of the Marine fire, the militia withdrew into the rubble of Beirut. From then on they adopted a more circumspect approach to the Marine snipers on the University building.

During the time that the Marines were holding their beleaguered positions in Beirut, American troops were involved in a small war on the Caribbean island of Grenada. On 25 October 1983 US Army and Marine troops landed on Grenada to suppress a Cuban-supported coup on the island. Resistance was fiercer than expected, especially from the Cuban troops stationed on Grenada, who were quick to react to the American assault. Rangers from the 75th Infantry Regiment found themselves under accurate mortar fire from a Cuban position guarding the Point Salinas airfield. A sniper team was sent ahead to deal with Cubans. Adopting a position which provided a clear field of fire, the Ranger snipers inflicted eighteen casualties on the mortar position, which collapsed as a result of this deadly, unseen fire.[5]

As the US forces pushed forward to deal with the last centres of resistance they continued to meet with determined fire from dug-in Cuban troops. One platoon of Marines was held up by a Cuban sniper, armed with a Soviet SVD sniper rifle, as they attempted to cross a piece of open ground two hundred yards wide. The pressure was on the lieutenant commanding the Marines to keep the assault on schedule, but to cross the open ground would lead to casualties. The call, 'Sniper up!' went down the line, and the sniper team acting in support of the Marines assessed the situation. The sniper and his observer scanned the most likely Cuban sniper positions. The sniper said to his observer:

'Look down along the far side, where the old shed leans against the house. Now, if I was over there, I just might slip into that shadow, where I'd be out of sight and sort of protected. Then I'd put a round right up this way.'

'He just might try that. It looks perfect from there.'

'Easy shot if he crawled in there.' The sniper shifted a little and laid his cross hairs on the shadowed position.

'Hell!' A bullet sliced through the foliage. 'He's shooting back this way.' The sniper waited with outward patience. The observer said, 'He's down near there. I got a glimpse of him moving.'

The sniper's breath caught and held. The observer started to say, 'Hell, there he is . . .'

Crack! The Remington [M40A1] bucked a little in recoil. The sniper was already working the bolt.

Eyes glued to his binoculars the observer held on the darkened shadow. He felt the sniper again seek his target – just as a long rifle barrel poked into sight and hung for a short instant before the entire rifle slid into view, to tumble and slide to rest on a slant of dirt rubble.

A fatigue arm flopped into sunlight and lay motionless. Instantly the Marine sniper fired, and the arm leaped bonelessly before its now broken shape lay stilled.

'Got him.' The observer kept his binoculars working. The sniper had slid down a little. He thumbed a pair of live rounds into the rifle's magazine and slid the bolt closed.

'Signal the Lieutenant that we got the guy.'[6]

The general satisfaction felt by the US Marines with their M40A1 rifle only served to underline the misgivings felt by US Army snipers towards the M21, which by the 1980s was showing signs of increasing wear. The dissatisfaction was especially strong among special forces and airborne troops, who had a reputation for promoting sniping and found the M21 unable to hold its zero during airborne operations, as well as lacking in long-range accuracy.

After extensive testing, the US Army turned its back on a semi-automatic weapon in favour of a bolt-action rifle, utilizing the same Remington action as the M40A1 (assembled by the Marines, the M40A1 was produced in too few numbers to meet Army demand). Designated the M24 SWS (Sniper Weapon System)[7] the receiver is bedded into a composite synthetic stock with an adjustable butt plate. The rifle has an integral four-round magazine (plus one in the chamber) and an optional bipod. An interesting feature of the M24 specification is its ability for conversion to .300 Winchester Magnum ammunition for 1000-metre-plus shooting. The telescopic sight, the 10-power Leupold Ultra M3, was designed for the M24 and is capable of rapid removal and re-installation without affecting the zero. The first M24s entered US Army service in 1987 but because of high manufacturing costs it has not completely replaced the M21, which remains in use for back-up and lower echelon sniping.

During the 1989 invasion of Panama, US forces armed with M21s were chagrined to find M24s in an arms cache hidden by General Manuel Noriega's Panama Defence Force (PDF). Despite this, Sergeant William Lucas had an opportunity to use his old M21 to good effect during the Panama operation. A veteran sniper, with thirty-eight kills chalked up from his time in Vietnam, Lucas was able to redirect his talents in the urban jungle of Panama City. As the Americans began the task of tracking down Noriega, troops of the 82nd Airborne Division found themselves being harassed by snipers. Sergeant Lucas and his observer took up position on the fifteenth floor of the Marriot Hotel to eliminate a PDF sniper who was holed-up in a high-rise block 750 metres away. Journalist Bobby Feibel provided this account of the counter-sniper engagement, as US soldiers located the enemy sniper:

A member of Noriega's Panama Defence Force, dressed in black, makes the mistake of presenting a silhouette. Lucas sights in, takes the slack up in the trigger. His breathing slows, then stops. The silhouette disappears. Lucas waits.

The shade on the targeted window moves and the spotter catches the gleam of sunlight on a barrel.

'He's setting up for a shot,' the spotter says quickly.

Lucas peers through the adjustable ranging telescope. He exhales, takes up the slack in the trigger and squeezes. The rifle booms, echoing in the room.

'He's down. I saw him fall,' says the spotter, eyes still glued to his binos. Later, Lucas sees a woman open the door to the room he fired into. She walks around the apartment and finally

213

out onto the balcony. Her hands fly to cover her mouth and she leaves quickly. Lucas says: 'Judging by her reaction, I'm pretty sure we either killed him or hit him hard.'[8]

The operation in Panama was only an advanced police mission, compared to the next major deployment of American troops in the Gulf War of 1990-91. This was a full-scale military deployment involving all the services. Following the extensive softening-up by the Coalition air forces, the Iraqi Army was in no fit state to face the land campaign, which led to complete military collapse after a hundred hours of combat. Before the final Blitzkrieg phase of the war, Coalition snipers were active in eliminating key targets as part of the ground forces' contribution to harassing the Iraqis. Once the Coalition armoured units broke through the Iraqi lines, the snipers were able to continue playing a useful role, especially since many of them were armed with lightweight .50-calibre sniper rifles, capable of penetrating light armour.

During the 1980s American gun makers began to meet the long-awaited demand for a .50-calibre weapon that was man portable and matched the accuracy requirements of a conventional 7.62mm sniper rifle.[9] Using high-power telescopic sights these powerful weapons can fire a 709-grain bullet to ranges in excess of a mile with high accuracy. The McMillan company brought out a five-round bolt-action rifle in 1987, weighing only 25lb. Other heavy sniping rifles include the AMAC .50-calibre sniping rifle, a single-shot bolt-action weapon, and the Barret M82A1, a semi-automatic rifle capable of holding five- or ten-round detachable magazines. In 1990, Barret produced the M90, a bullpup design (where the receiver and magazine are behind the trigger and grip), with an overall length of 45 inches.

The Barret M82A1 was used in the Gulf War; a hundred rifles were rushed to the Marine Corps in time to see action in the desert. In one engagement, Sergeant Kenneth Terry of 3rd Battalion, 1st Marines, hit and knocked out an Iraqi BMP armoured personnel carrier with two armour-piercing incendiary rounds at a range of 1100 metres. At the loss of the Iraqi vehicle the other two BMPs in the patrol promptly surrendered to the American forces.[10] Confirmed kills were made against human targets at ranges of 1800 metres, and one unconfirmed report suggests a kill at a range of 2200 metres. An unconventional task allotted to .50-calibre snipers lay in shooting and disposing of the anti-personnel mines, sown so profusely by the Iraqis in the desert areas of Kuwait.

Well trained and equipped, the Coalition forces were operating in an excellent battleground for sniping. When found, Iraqi targets were swiftly despatched, whether by US Marine M40A1s, US Army M24s/M21s or British L96A1s. The Soviet-trained Iraqi snipers were not in the same league as their opponents, and wisely they kept out of the battle. Besides better training and arms, the Coalition snipers had vastly superior morale: they were confident of victory. One US Marine summed up the prevailing mood: 'The more Iraqis I kill, the less of my buddies get killed.' Typical of the black humour of combat troops was the verse sung by Marine snipers to the tune 'Winter Wonderland':

As my M40 fires,
His life will expire.
A shot to the head,
My target is dead.
Walking in sniper wonderland.[11]

The Gulf posed special challenges for the snipers operating in a desert environment. Apart from the obvious problems of dust and water shortages, the featureless desert made the judgement of range and wind speed difficult, even if optical and laser range finders (where available) were a means of overcoming this problem. Because of the long ranges, observers dispensed with their M16s in favour of M14/M21s as sniper-team support weapons.[12]

The mountains of Afghanistan, where in the 1980s the Soviet Union waged a remorseless guerrilla war against the Afghan Mujahadeen, was another hostile environment which provided snipers with new opportunities. Trained and equipped to fight a European-style war, Soviet troops experienced immediate difficulties in Afghanistan, where the open terrain encouraged long-range shooting. Used to operating from within a BTR or BMP vehicle, the Soviet soldier was reluctant to leave its safety to engage the Afghan guerrillas in a direct firefight.

Shooting standards on both sides were generally low, but the Mujahadeen possessed at least one advantage in equipment. Whereas the Soviet Kalashnikov assault rifles were ineffective over 300 metres, the Afghans, many armed with old British Lee-Enfields or commercially acquired full-power rifles, were able to engage the enemy at ranges of up to 800 metres. As the war progressed the Afghans were able to procure more advanced weapons from the West, including several types of sniper rifle, which were highly effective against the road-bound Soviet troops.

The Soviet answer was to triple its sniper capability, so that instead of a single sniper being attached to a rifle squad, a complete sniper squad/section was added to every company in the field. This gave the company commander a long-range precision fire capability; their prime function was to act as a counter-sniper force, able to winkle out Afghan guerrillas hidden in crags overlooking Soviet positions.

The Soviet sniper weapon in Afghanistan was the SVD (*Snayperskaya Vintovka Dragunova*, or Dragunov Sniper Rifle), which fired the full-powered 7.62×54mm round used in the old Mosin-Nagant rifle.[13] Designed by Yevgeny Dragunov, the SVD has certain similarities to the AK–47, not least in its external design, but the internal parts of the rifle differ in many ways. A semi-automatic weapon the SVD has a ten-round detachable magazine, a skeleton stock, and a combination flash-suppressor and compensator to reduce muzzle jump and flash. Each rifle comes with a 4-power PSO–1 telescopic sight, which has an integral range-finder that can be lit up in poor light. The scope has its own infra-red detector which can locate enemy sources or be used passively with external IR light projectors.

The SVD cannot be compared to a Western bolt-action sniper rifle: its semi-automatic action compromises accuracy, as does its thin barrel (which could not realistically be free-floated), and its scope has only limited powers of magnification. The SVD is an accurate rifle out to a maximum range of 600 metres; at 800 metres it has only a 50 per cent chance of hitting a stationary, man-sized target. Despite these limitations the SVD performs well in its basic role, to act as a long-range sharpshooter's weapon.

The SVD entered service in the Red Army in 1967; since then it has been exported throughout the old Warsaw Pact and to other Soviet client nations. The Chinese have made copies of the SVD, as did the Army of the former Yugoslavia, under the designation M76. The M76 is a high-quality copy of the SVD, but the Yugoslavs, with an eye on the export market, manufactured the rifle in 7.92mm, NATO 7.62×51mm and Soviet 7.62×54mm calibres. As the M76 was produced in large quantities, intended to equip a sniper in every squad/section, many of them have been used in the conflict that followed the break-up of Yugoslavia into its separate ethnic and religious groupings.

Since war broke out in the former Yugoslavia in 1992, the term 'sniping' has become a coverall name for almost any form of rifle shooting, aimed or otherwise. As the civilian population is both an easy (and important) target for the military forces on all sides, they have borne the brunt of much of this so-called sniping. Any soldier with the most basic weapons training can loose off a few rounds at a group of women and children clustered around a water tap and reasonably expect to cause some casualties. This sporadic hit-and-run type of warfare is typical of much of the fighting in Yugoslavia. Shooting at the civilian population – whether by rifle or machine-gun – is facilitated by topography and ethnic division, so that, for example, a town held by one group is surrounded by high ground under the control of another. The latter are able to fire into the town at will; casualties will inevitably occur.

Alongside this opportunist shooting, the armed forces fighting each other deploy snipers in a conventional military manner. There was a strong hunting tradition in Yugoslavia, and the Yugoslav Army fostered rifle shooting. In the confused situation that characterizes the war in the former Yugoslavia, all sides have drawn upon outside sources of supply. This has been especially true of the Croats and Bosnians, who lost out to the Serbs in the military equipment carve-up of the old Yugoslav National Army. In sniping terms, this ensures that there is a profusion of different weapons, including M76s and commercial models bought from Germany and Austria, or even the United States. Some of the equipment is first-rate and includes advanced optical equipment and laser rangefinders (one journalist with military experience remembers going up to his rocket-hit hotel room in Sarajevo as dusk fell, only to encounter the disconcerting sight of a red dot from a laser range finder flitting round the wrecked bedroom).

In Sarajevo both Muslim and Serb snipers have had the opportunity to put theory into practice on a grand scale. While the Serbs possess the advantage of terrain, holding most of the high ground circling the city, the Bosnian Muslims have become highly proficient snipers, inflicting casualties on the Serb forces. To counter this threat, the Serbs formed special anti-sniper units. A member of one of these units describes the war of snipers in the battered streets of Sarajevo:

> It's an exhausting task which calls for nerves of steel and endless patience, especially when we find ourselves pitted against snipers. A few months ago, a friend of mine was shot dead as he was running down the street. The bullet that killed him could have been fired from any of several three-storey buildings. I spent five days scouting around for clues but the sniper made no mistakes, he kept on moving about, changing fire positions constantly. Late one day, however, I caught a fleeting glimpse of a silhouette in my scope. At long last I had located one of his hide-outs. The next day, the man ventured out of the shadows when a fusillade broke out in the neighbourhood. Never suspecting that he had been detected, he stepped out into the light close to a window. I fired only one bullet.[14]

The confused military situation on the ground has been complicated further by the influx of mercenary troops, a mix of starry-eyed adventurers, who soon discovered the savage reality of Balkan-style warfare, and ex-soldiers plying their trade. The Croatian side has attracted most of the Western mercenaries (mainly ex-British forces and French Foreign Legion), employed to train local forces and stiffen the line in trouble spots. A number of these men are trained snipers whose expertise is invaluable at the front. An account of a sniping operation was written by an ex-Australian soldier, 'Skippy' Hampstead, fighting on the Croatian side. Fighting against Serb forces in the Mostar region of Bosnia, Hampstead and a British mercenary were engaged in counter-sniper operations. They were armed with a Yugoslav M76 and an Austrian 7.62mm Steyr SSG sniper rifle. Hampstead describes a successful sniper mission:

> On our return to Mostar we met up with some residential snipers who took us to some excellent locations. From one building I had a clear view of a crossroads across the river, and it was here that my success rate went up considerably. I scored my best shot of the war.
>
> We had been there since about 8 a.m. and we'd had plenty of trade. They had absolutely no idea where we were, so there was no return fire. About 11 o'clock a car appeared, driving very slowly towards me. 'A gift from heaven,' I thought. I aimed carefully at the driver, allowed for movement, and a metallic click of an empty chamber was the result. By the time I had reloaded, the car was gone. I was furious with myself. I had seen three men inside the car; it had been a wasted opportunity.
>
> One cigarette and half an hour later the car reappeared, and turned away from me at the crossroads. Squeeze – *Crack!* There was a small orange flash, a boom, followed by an even bigger flash and a louder boom as the whole car blew up in a muffled explosion. I couldn't believe it. One shot equalled three enemy soldiers, three weapons, possibly three pistols and one vehicle.[15]

After this success the Serbs were more careful in their movements and did not present further targets, and the two snipers decided it would be wise to withdraw from the position. Hampstead's final comment underlines the economy of sniping, how a single rifleman can be as effective as an infantry squad.

The fighting in the former Yugoslavia has once again demonstrated the continuing opportunities open to snipers. There are few tactical situations where a knowledge of sniping and suitable equipment cannot be put to good use.

CHAPTER FOURTEEN

THE SNIPER TODAY

At the end of the Second World War, the sniper's position within most armed forces was tenuous. Over the last fifty years, however, the sniper has slowly worked his way back into the heart of the military machine, and today it is generally recognized that he is, in military jargon, 'a cost-effective force multiplier'. In other words, the sniper is relatively cheap to train, equip and deploy, yet he provides the battalion or company commander with a variety of tactical options, increasing his overall firepower at ranges far beyond those of other infantry.

While the sniper is now appreciated within the armed forces, the question of deployment remains. In America and Britain, snipers are organized at battalion level (although US Ranger units deploy their snipers with the company) but because of their small size (eight to sixteen men) they remain a small cog within the battalion structure. The sniper section, considered too small to exist in its own right, is usually tacked on to some other, larger unit.

In the Royal Marines, snipers are part of the reconnaissance troop.[1] This is in many ways a sensible arrangement – snipers are highly trained in reconnaissance work – but their roles are ultimately very different. A reconnaissance soldier must avoid enemy contact at all costs, but a sniper's prime mission is to engage the enemy. As a consequence, the snipers within the recce troop tend to find themselves overwhelmed by a larger group whose tactical outlook is to minimize contact.

This is not an isolated phenomenon. Snipers in the US Marines are termed scout snipers, reflecting their reconnaissance function, and today are part of the STA (Surveillance and Target Acquisition) platoon. The end result, however, is that sniping

220

suffers, an issue raised by Marine officer Captain Edward D. Daniel, who writes: 'The most critical problem is that many battalions concentrate almost solely on the scouting or close-in reconnaissance role assigned their scout snipers, at the expense of the sniping role.'[2]

It would be foolish to argue that snipers should be exempt from the realities of working within a military organization, where compromise of some sort cannot be avoided. Yet, equally, snipers should not be completely subsumed within another unit. Rather they should maintain a semi-independent status within the battalion headquarters. As a result, battalion and company commanders will be more aware of their potential, and where and when they might be most effectively used to support the parent unit.

Sniper instructors may bemoan shortages of resources, but the standard of basic sniper training is generally high. More problematic is an awareness of the need for continuation training, to ensure that standards are maintained, and to increase the sniper's knowledge and abilities. The latter aspect is important, for a six- or eight-week course can only instil the rudiments of the subject, and a sniper requires opportunities to expand and refine his skills. Ideally, continuation training should allow the sniper to operate in different situations – in urban environments, in deserts and over snow-bound terrain, for example – each of which requires different sniping responses. The answer is to match resources to sniper requirements.

The sniper also needs to keep abreast of the latest technological advances which affect the conduct of his mission. Ever since the Vietnam War night-vision devices have become a standard part of military operations. To be able to fire at night is an obvious advantage to a sniper, but paradoxically, these advances have also worked against him. Because of the limitations of night sights – particularly in depth perception – ranges are brought down to little more than four hundred yards. Thus the long-range advantage of the sniper's rifle is greatly reduced as ranges come within that of the standard infantryman's rifle. And as the use of night-vision devices has become more widespread, so the sniper is more vulnerable to detection. This risk has grown as a

221

result of parallel advances in ground radar and thermal imaging techniques. The protective cloak of darkness is becoming increasingly threadbare.

Not all is lost to the sniper, however, as his superior skills in fieldcraft and marksmanship still give him the edge over the rank and file infantryman. Recent developments in night-vision technology have also helped the sniper. Progressing beyond standard, second-generation night sights, such as the AN/PVS-4, are yet more sophisticated aids to night shooting. The Litton M938, with a 6-power lens, helps restore the sniper's long-range advantage, enabling accurate shooting beyond six hundred yards. The Simrad KN-200 is a light-intensifying device that is clamped directly on to the rifle's ordinary optical scope, so that the sniper is able to fire out to ranges of eight hundred yards by looking through his own telescopic sight. The Simrad can then be taken off in daylight without affecting the rifle's zero. When fitted to a .50-calibre sniping rifle, this scope has been able to hit targets on a moonless night at ranges of up to a thousand yards.[3]

The rapid progress in micro-chip technology has filtered down to the sniper. Navigation for sniper teams operating deep behind enemy lines – mainly special forces' missions – has been transformed by hand-held satellite navigation devices such as the Magellan Global Positioning System. They enable a soldier to fix his position anywhere in the world without reference to external features, except an overhead satellite. These devices were invaluable in the Gulf conflict, enabling troops to conduct missions over featureless desert without the almost inevitable concomitant of getting lost.

Radios have become lighter and more reliable. In the past, the sniper team was on its own after it left friendly lines. Since the 1960s and the advent of genuinely portable radios, it has been possible for the sniper team to relay vital intelligence directly back to base, or if in a dangerous situation, to call for rapid backup. The comparatively primitive radios of the Vietnam era have been replaced by a new generation of lightweight devices – the PRC–77 and increasingly the PRC–113 – which provide the isolated sniper team with a secure communications network. The advantages of good radio communications work both ways, allowing the sniper squad/section leader to direct his snipers far more effectively. The sniper team can utilize its greatest tactical asset – freedom of action – and yet remain under the control of the parent infantry unit.

Rifle suppressors have also benefited from recent technological progress. Although used on a limited scale in Vietnam, suppressed rifles have not been adopted for conventional military sniping because their ability to reduce noise does not overcome their accuracy disadvantage. Quietly missing the target is no substitute for a kill, no matter what the sound level.

There are two types of suppressed sniper rifle. First, those rifles that fire standard supersonic ammunition with a suppressor attached to the muzzle, where only the noise of the muzzle blast is reduced, but not the supersonic crack of the bullet's flight. The absence of muzzle blast confuses the enemy trying to locate the sniper, as the crack alone is hard to pinpoint. Second, those rifles that use a suppressor and fire subsonic ammunition which removes the crack altogether, but dramatically reduces accuracy to the degree that ranges of a hundred yards are normally an absolute maximum.

Rifle suppressors remain within the field of special forces operations, despite design improvements. Several rifle manufacturers have brought out suppressed sniper rifles which are more compact than the unwieldy adaptations of former years. Among these are the McMillan M89, the Steyr SSG P–IV 'Urban Sniper' and the Vaime SSR Mk 1/2, used by the SAS, where the suppressor is an integral part of the rifle. The advantage of this generation of rifles is that they have been designed to be suppressed. Arguably, the most specialized suppressed sniper rifle is the De Lisle Mk 4, a tailor-made weapon based on the Remington 700 action, firing 7.62mm rounds in either standard or subsonic combinations. The rifle utilizes a heavy match barrel with integral suppressor, which improves accuracy, and using subsonic ammunition enables targets to be hit at ranges of over two hundred metres.[4]

Looking towards the future in sniping technology, work is progressing on the development of a lightweight integrated sight that will combine a telescopic sight with a night sight and an infra-red rangefinder and thermal imager. This would be of great use to a sniper, but like so many technological advances it has its disadvantages. Snipers who rely too much on technology will lose the hard-earned practical skills which are the basis of their discipline. As soon as anything goes wrong with their various devices (as it always will), they will be rendered virtually useless. More insidious still, once snipers lose their feel for the physical nature of their calling they will lose their feel for sniping. Carlos Hathcock was armed with a rifle whose design dated back to the turn of the century, and his skills in fieldcraft and marksmanship were based on knowledge older still. It is unlikely that laser rangefinders would have made him a better sniper.

A second, crucial disadvantage of an increasing reliance on technical devices is one of simple cost. The more the sniper is equipped with electronic devices as a matter of course, then, with a few exceptions, the less will he be cost effective. And cost-effect is one of the essential premises behind sniping, especially in a world where defence budgets are being asked to pay for the 'peace dividend'. This is not to deny the worth of technological progress in sniping; it is a question of achieving balance, to be able to judge just how effective a piece of equipment is for the sniper's mission.

During the 1980s rifle design saw the consolidation of the heavy-barrelled, bolt-action weapon, utilizing a synthetic stock, and fitted with a high-magnification fixed-power scope. The US Army joined the US Marines and the British Army and Royal Marines in using bolt-action rifles – a testament to the efficacy of this weapon system.

In spite of assertions made by some snipers and gun makers, the semi-automatic rifle does not have the accuracy of a bolt-action rifle. Only Heckler and Koch's semi-automatic PSG–1 can make serious claims to a similar level of accuracy, but this rifle is very heavy (17.8lb unloaded) and very expensive. Designed for police counter-sniper work it is a fine piece of engineering, and the company has brought out a simplified military version, the MSG–90. But whether such a rifle is better for a military sniper than one with a conventional bolt action remains a matter of debate.

The obvious advantage of the semi-automatic over the bolt-action rifle is its ability to fire successive shots quickly; snipers do not always hit home with the first round, and often targets comprise groups of enemy troops who can be caught by swiftly aimed fire. Proponents of the semi-automatic sniper rifle argue that these benefits outweigh a loss in long-range accuracy. Yet, in turn, those who favour the bolt action point to additional problems with semi-automatics. The Vietnam sniper Major Jim Land played an important role in developing modern US Marine sniping, based around the bolt-action M40A1. He puts forward this criticism of snipers armed with semi-automatics: 'The sniper tends to get involved in the firefight, which is not his mission, and when this happens his mission suffers.' Land also points out that with a semi-automatic, the cartridge case is flung out of the rifle during extraction, a potentially fatal action:

> In Vietnam, we spotted more enemy troops from the flash of the sun off the flying brass from their Kalashnikov assault rifles than from any other cause; for this reason I prefer a bolt action which will allow the sniper to slowly extract the spent case and put it back in his pocket. Brass from an autoloader [semi-automatic] must be policed up if you don't want to give away a hide you might want to use again.[5]

Although the case for either system cannot be conclusively proved either way, the argument for bolt-action sniper rifles remains the stronger. Yet the differences between bolt action and semi-automatic should not obscure the fact that the modern sniper rifle is a superb weapon, the product of years of craftsmanship by gun makers meeting the demands of civilian and military marksmen. In contrast to most other weapon systems, the design of the sniper rifle has remained much the same throughout the century; with the exception of the use of synthetic stocks, there have only been small modifications to the basic model. The most interesting new sniper rifle of recent years has been the Walther WA2000, which incorporates a highly efficient muzzle brake, improved magazine and a bullpup configuration, where the trigger assembly is forward of the magazine and receiver group.

With an overall length of just under a metre, the WA2000 is by far the most compact sniper rifle on the market, even if its design represents an aesthetic nightmare to the traditionalist.

Whatever the future for sniper rifle design, it is over-shadowed by the debate over ammunition performance. This is the bottleneck in the progress of long-range shooting. Ammunition in 7.62×51mm calibre is not particularly well suited as a sniper round; it lacks penetration and does not perform well at ranges beyond 700 to 800 yards. With hindsight this choice of round as the NATO standard was not a particular success. It was too powerful to be an effective round for automatic/assault rifles, and within a decade of its introduction, plans were afoot to replace it with the smaller calibre, reduced-load 5.56mm cartridge. The 7.62×51mm round is a reasonable rifle cartridge, but like many compromise options, it is not sufficiently effective for sniping.

The 7.62×51mm M118 Special Ball ammunition has been the subject of much criticism from snipers, despite its obvious superiority over standard M80 ball ammunition. As early as 1973, doubts were expressed over the quality of the 173-grain bullet produced at the Lake City Arsenal. During the 1980s protests grew in strength, especially as it was claimed that quality control had fallen further. Civilian shooters preferred Federal's .308M match ammunition, and by 1990 military snipers were relegating the M118 round for practice use, replacing it with M852 match ammunition with the Sierra 168-grain MatchKing bullet used in the Federal round. (Problems for combat use remain, however, due to the fact that the Sierra bullet is technically a hollow point – for manufacturing reasons only – and thus proscribed by the Hague Convention.)[6]

Although an improved bullet would be welcomed, the 7.62×51mm cartridge is not sufficiently powerful, and snipers would prefer to see the adoption of a new ammunition type. An obvious contender is the tried and tested .300 Winchester Magnum round, which has been particularly successful in 1000-yard shooting at the US National Rifle Association's Camp Perry competitions. A Federal Premium .300 Winchester Magnum round firing a 200-grain boat-tail bullet has a muzzle velocity of 2830fps and an energy level of 3560 ft/lb, which compares favourably against the Federal .308 match round, firing a 168-grain boat-tail bullet with a muzzle velocity of 2600fps and an energy level of 2520 ft/lb. The energy of the Winchester Magnum declines slowly, so that between 600 and 1000 yards it is almost double that of the standard .308/7.62mm Federal bullet.[7]

The main objection to the adoption of the magnum round is still the old dislike of adding another ammunition type to the logistical system. But the fact that Britain's L96A1 and the US Army's M24 sniper rifles (among others) have been built with a capacity to take the .300 Winchester Magnum round, is suggestive of an eventual move away from the straitjacket of 7.62×51mm ammunition.

Another development in magnum ammunition has been the introduction of the .338in (8.6mm) Lapua Magnum round, firing a 250-grain boat-tailed bullet, and developed for European military sniping (the versatile L96A1 can be rechambered for this calibre). The idea behind the 'super' magnum concept is to provide the sniper with an improved, long-range capability that includes the penetration of buildings, motor vehicles and aircraft, while at the same time enabling the sniper to carry a conventional bolt-action rifle which is manoeuvrable and light enough for standard operations. Medium/heavy sniping is still at the experimental stage, but work is in progress to introduce ammunition that can be used in an anti-personnel capacity at ranges of 600 to 1500 metres. This is a development of great interest to snipers and will have ready application for special forces units such as the SAS or US Navy SEALs.

Progress in .50-calibre sniping continues. The 12.7×99mm round, firing a 709-grain bullet, dominates heavy-calibre shooting, and even though there have been experiments in other, even larger calibres, it is likely that .5in will remain as the standard for heavy sniping. For reasons of weight and size the .50-calibre weapon can never replace the 7.62mm (or perhaps .300) sniper rifle, but it is a valuable resource, whose armour-piercing capabilities open up new opportunities to the sniper. Thus, for example, the .50-calibre SLAP (Saboted Light Armour Penetrator) round can cut through 19mm of steel plate at a range of 1200 metres, a level of performance that enables snipers to engage light armour.

The greatest changes taking place in sniping lie in the field of target acquisition. Ever since British snipers were issued with armour-piercing bullets to destroy the breeches of German machine-guns during the First World War, snipers have looked for a way to expand the range of targets beyond enemy soldiers. The development of .50-calibre and magnum sniping rounds has enormously increased these possibilities.

This trend has been facilitated by the increasing sophistication of modern warfare, placing a profusion of vulnerable, high-value targets within the sniper's sights. One sniper authority even suggests this may lead to a transformation in the sniper's mission: 'Given the key importance of several electronics devices (and particularly of C^3 systems) it would not be surprising if target priorities are eventually changed, the sniper being mainly tasked not with the elimination of selected individuals but rather with the neutralization of equipment such as microwave antennae, power generators and transformers, search guidance or C^2, and so on.'[8]

Yet, the role of the sniper as a hunter of men will undoubtedly continue, and as this task is by far the hardest facing a sniper, it will remain the core discipline of sniper training. The increase in the number of available targets merely acts to strengthen the sniper's position within the military organization.

Another recent trend lies in issuing a marksman in each infantry platoon with an improved assault rifle fitted with a low-powered telescopic sight. At a conventional military level, the concept of a 5.56mm sniper must remain a contradiction in terms, but there is no reason why a platoon should not have an improved sharpshooting capacity. The requirement for such a sharpshooter will vary from army to army. In the British armed forces, for example, every rifle is equipped with a 4-power SUSAT sight, and for all the SA80's faults, it is an accurate assault rifle which, combined with a generally high level of marksmanship, reduces the requirement for a dedicated sharpshooter. Within the British infantry section, the long-range fire role is undertaken by the heavy-barrelled Light Support Weapon, which, despite its

5.56mm calibre, has an increased range. In other armies, however, a sharpshooter may well be a useful addition to the section or platoon, although he can never take the place of a true sniper.

Since the end of the Cold War, the armed forces of the West have had to prepare to fight smaller and more varied wars. Previously, the massed tank battle to be fought in Central Europe directed tactical requirements along quite specific avenues. The emphasis was on armour and defeating armour, and snipers were left on the periphery. Since then, the focus of attention has changed to more flexible operations, where the sniper has a more obvious role. The future for sniping looks optimistic. New technology, despite its limitations, is opening up the sniper's tactical outlook, and it only remains for infantry commanders to have the knowledge and imagination to use this most valuable of assets to the best advantage.

PART FOUR

WEAPONS OF WAR

'The average rounds expended per kill with the M16 in Vietnam was 50,000. Snipers averaged 1.3 rounds. The cost difference was $2300 v. 27 cents.'

Sign at US Marine Corps Scout Sniper School,
Quantico, Virginia

CHAPTER FIFTEEN

RIFLES

The rifle is the sniper's most precious tool: the better the rifle the better the sniper. A high-quality rifle enables a sniper to perform his essential mission of long-range precision shooting. To build a sniper rifle, the manufacturer must use only the finest materials, ensure that tolerances are fined down to a minimum, and impose a draconian level of quality control. Many items will fail to pass muster as a matter of course, so that, for example, during testing of the US Army M21 rifle, 25 to 30 per cent of rifles were rejected.[1] This inevitably makes sniper rifles expensive to manufacture, especially as production runs are restricted to a few thousand examples; there are no economies of scale to recoup initial costs.

Four main elements constitute a sniper rifle: the receiver and action; the barrel; the stock; and the sights and mount. Each part must work in harmony with the others if the rifle is to be effective.

The receiver is subject to great stresses when a shot is fired, and manufacturers attempt to make it as strong and rigid as possible. This is a major reason why the bolt action is chosen in preference to a semi-automatic action. When the bolt – especially the Mauser front-locking type – is closed, it reinforces the rigidity of the whole receiver, in a manner not possible in a semi-automatic with its many moving parts and pressed steel receiver. The trigger action must be as smooth as possible, with a trigger pull of around 3–4lb, and a quick-firing pin fall (or lock time), so that the time between pulling the trigger and the bullet travelling up the barrel is as short as possible.

The barrel of a sniper rifle is made from high-quality carbon or stainless steel. The barrel is similar to a tuning fork when fired, and if accuracy is to be maintained the vibration of the barrel must be even and uniform. Consequently, the barrel is free-floated, that is, nothing touches the barrel – the fore-end of the stock, for example – except where it is screwed into the receiver. Accuracy is also affected by barrel width; the thinner the barrel the greater the amplitude of the vibration and the less the accuracy. To minimize vibration the sniper rifle is normally fitted with a heavy barrel possessing a diameter of up to an inch (25mm) at the muzzle. To reduce the sniper's visibility when firing a shot and to diminish recoil, it is common to attach a combined flash hider and muzzle brake at the end of the barrel.

The stock must be strong and not warp through changes in humidity and temperature, and to minimize this possibility a wooden stock is usually impregnated with an epoxy resin. To make a good match between receiver and stock, the receiver is glass fibre bedded into the stock to ensure a firm bond between steel and wood. Increasingly, however, wooden stocks are being replaced by synthetic types which are stronger and not affected by climatic changes. Thus, while the US Marines' M40 utilized a conventional wooden stock, its M40A1 replacement has a stock made from fibreglass. Most synthetic stocks have an adjustable butt stock (in length and/or height) to allow for differences in human morphology. This enables the sniper to gain a good sight picture without having to strain his neck or arms. As a further aid to accurate shooting, most modern sniper rifles are fitted with an adjustable (and detachable) bipod.

The telescopic sight is the most complex part of the sniper weapon system, but fortunately for the sniper it is a self-contained unit which does not require field maintenance. The sight must be easy to attach to and take off the rifle, hence the need for a solid mount which does not affect the working of the rifle and can take different types of sight.

Telescopic sights have improved enormously over the last two decades, and are now far more reliable, have improved luminosity and provide greater magnification. Yet the basic problem of reconciling high magnification with luminosity and a reasonable field of view remains. The greater the magnification, the less light can get through the scope to the sniper's eye and the smaller the field of view (thereby restricting target acquisition). A compromise is made between the various elements, so that in most military telescopes, magnification is between 6- and 10-power, with a relatively large objective lens (which collects the light) of around 40mm. (Lens specifications are usually expressed as a function of their magnification to objective lens size, so that a designation of 6×42 means a 6-power scope with an objective lens 42mm in diameter.)

Military sights will normally have some sort of ranging device within the sight reticle. As part of the rigours combat inflicts upon a telescopic sight mounted on a sniper's rifle (in contrast to a scope used only for target shooting), it must be robust, waterproof and reasonably lightweight.

Although the conventional telescopic sight remains the standard sighting device used by the sniper, the development of night-vision aids has made a dramatic change to his mode of operation. The introduction of the most advanced night sights, such as the Litton and the Simrad, has been of considerable advantage to the sniper, making possible accurate shooting at ranges greater than that of the assault rifle. Yet they still cannot compare to the optical sight and night operations remain, for the time being at least, an adjunct to daylight sniping.

The purpose-built military sniper rifle is a surprisingly new phenomenon. Sniper rifles were traditionally modifications of existing weapons, derived from two quite separate sources. The first was the improved military rifle, an uprated version of the armament carried by the ordinary soldier, on to which was clamped a telescopic sight. As the service rifle was usually a reasonably accurate weapon, so a reasonably accurate sniper rifle could be developed, and yet it inevitably failed to meet the special requirements of a sniper. The second source was the sporting rifle, a hunting weapon taken up by snipers because it was generally more accurate than a service rifle. But this choice was also less than satisfactory, because the weapon was never designed for the rigours of military life, and sporting and military calibres were often dissimilar.

The snipers of most nations tended to use uprated service rifles, although some – notably the United States – adopted both, so that during the Vietnam War, the US Army employed an accurized M14 service rifle, while US Marine snipers were armed with modified Winchester and Remington sporting rifles.

The first real military sniping weapon was the Soviet SVD and, although similar to the Kalashnikov series of assault rifles, it was designed exclusively for sniper use. The West was slow to follow the Soviet lead and it was only in the late 1970s and the 1980s – as the adoption of small-calibre assault rifles made sniper versions impracticable – that dedicated sniper rifles were introduced. Since then, however, they have been produced in profusion, manufacturers vying with each other to improve on existing designs. As a consequence today's sniper rifle is an extraordinary product, a combination of the tried and tested incorporating the latest technological advances.

Some armed forces have followed a broadly evolutionary approach to sniper armament; the US Marines moved gradually from the Winchester and Remington rifles to the Remington/M40A1. Others, by contrast, have adopted a more revolutionary attitude. The British Army, for example, had clung to uprated examples of the old Lee Enfield rifle until the mid-1980s, so that the introduction of the PM series marked a complete transformation of British sniping armament. Whatever approach has been chosen, the modern sniper rifle is more than just a synthesis of the accuracy of the sporting arm and the robustness of the service rifle, it is a unique weapon in its own right, pushing back the borders of military technology.

In the pages that follow many of the more important sniper rifles have been listed, although no attempt has been made to be comprehensive.[2] Also, this survey includes weapons that could not be called sniper rifles, but none the less are rifles that have been used extensively in sniper operations.

AUSTRIA

Steyr SSG 69 Sniper Rifle

Calibre: 7.62×51mm NATO
Operation: bolt action
Feed mechanism: 5-round rotary magazine, or 10-round box magazine
Weight: 8.6lb (3.9kg) empty; 10.1lb (4.6kg) with telescope
Length: 44.9in (1140mm)
Sights: 6-power Kahles ZF69 telescopic sight, plus emergency iron sights
Muzzle velocity: 2822fps (860m/s)

Developed for the Austrian Army, the SSG 69 (*Scharfschützen-gewehr* 69) is also used by several police forces. One of the more unusual features of this rifle is its five-round rotating spool magazine, housed within the rifle body. The stock is made from a synthetic material and the butt can be adjusted for length with the addition or removal of butt spacers. The standard Kahles ZF 69 sight is graduated for firing out to 800 metres, and can be replaced with infra-red and image-intensifying sights.

BELGIUM

Model 30–11 FN Sniper Rifle

Calibre: 7.62×51mm NATO
Operation: bolt action
Feed mechanism: 5- or 10-round detachable box magazine
Weight: 10.69lb (4.85kg) empty without telescope
Length: 44in (1117mm)
Sights: iron sights (anschutz aperture with adjustable dioptre) or 4×28 FN telescopic sight
Muzzle velocity: 2789fps (850m/s)

Intended for primarily police and other paramilitary units, the FN rifle is still a useful military sniping weapon. The walnut stock is adjustable both for height and length, and a bipod (from the MAG machine-gun) can be fitted to the fore-end of the stock. The FN telescopic sight is graduated from 100 to 600 metres.

CANADA

C3A1 Sniper Rifle

Calibre: 7.62×51mm NATO
Operation: bolt action
Feed mechanism: 6-round detachable box magazine
Weight: 13.89lb (6.3kg) empty with telescope
Length: 44.9in–47.7in (1140–1210mm)
Sights: 10-power Unertl telescopic sight

A development from the C3 (see Parker Hale Model 82 in UK section), the C3A1 incorporates features specified by the Canadian Army, including new rifling and a modified bolt handle that can be operated more easily by a sniper wearing gloves. A Parker Hale bipod is fitted to the stock, and the sights comprise the powerful Unertl scope used by the US Marines.

FINLAND

Sako TRG–21 Sniper Rifle

Calibre: 7.62×51mm NATO
Operation: bolt action
Feed: 10-round double-row box magazine
Weight: 11.69lb (5.3kg) empty without sights
Length: 47.3in (1200mm)
Sights: mount for telescopic/night sights (as required) and emergency iron sights

The tempered steel barrel of the TRG–21 is fitted with a detachable muzzle brake and flash hider. In line with recent thinking regarding the replacement of the 7.62×51mm round, the bolt and receiver have been designed to be able to accept a more powerful cartridge at a later date. The stock is manufactured from wood or synthetic materials, and features an open thumb hole, and can be adjusted for both height and length. A bipod is attached as standard.

Vaime SSR Mark 1 Silenced Sniper Rifle

Calibre: 7.62×51mm NATO
Operation: bolt action
Feed: 5-round box magazine
Weight: 9.04lb (4.1kg) empty without sights
Length: 46.5in (1180mm)
Sights: mount for telescopic/night sights (as required)

The Vaime has an integral suppressor/barrel assembly which makes it more effective and considerably more compact than a conventional rifle with suppressor attached. The stock is made from a synthetic material and comes with an adjustable bipod. The Vaime is a weapon for special forces applications, and as such is used by the SAS.

FRANCE

FR–F1 Sniper Rifle

Calibre: 7.5×54mm, or 7.62×51mm NATO
Operation: bolt action
Feed: 10-round detachable box magazine
Weight: 11.47lb (5.2kg) empty
Length: 44.8in (1138mm) without butt spacers
Sights: 4-power Model 53 telescopic sight, plus iron sights

The *Fusil à Répétition Modèle F1* was developed as a sniper rifle for the French Army. The action is based on the old Model 36. Noteworthy features include a pistol grip, padded cheek rest and two butt spacers for length adjustment and a bipod mounted at the rear of the fore-end of the wooden stock. The FR–F1 was manufactured in the traditional French 7.5mm calibre as well as the 7.62mm NATO standard.

FR–F2 Sniper Rifle

Calibre: 7.62×51mm NATO
Operation: bolt action
Feed: 10-round detachable box magazine

Weight: 11.47lb (5.2kg) empty
Length: 44.8in (1138mm)
Sights: various, including 6×42 and 3.2-power OB

An upgrade of the FR–F1, this rifle was taken into French Army service in 1984 and is now the standard sniper rifle. Improvements include a new plastic-coated metal fore-end, and a relocated bipod, yoke mounted just ahead of the receiver.

GERMANY

Mauser Model 98 Rifle

Gewehr 98

Calibre: 7.92×57mm
Operation: bolt action
Feed: 5-round integral box magazine
Weight: 9.26lb (4.2kg)
Length: 49.2in (1250mm)
Sights: iron sights
Muzzle velocity: 2100fps (640m/s)

Karabiner 98k

Calibre: 7.92×57mm
Operation: bolt action
Feed: 5-round integral box magazine
Weight: 8.6lb (3.89kg)
Length: 43.6in (1103mm)
Sights: iron sights
Muzzle velocity: 2477fps (755m/s)

The Mauser is one of the military classics, arguably the finest bolt-action rifle ever built. The original design dated back to 1888, and the revised Gewehr (rifle) 98 was the weapon that equipped the German Army during the First World War. Utilizing Mauser's famous forward-lug locking system, it was an accurate rifle that lent itself to sniping with the addition of telescopic sights.

Introduced in 1935, the Karabiner 98k was a cut-down version of the Gewehr 98, and was the main service armament of the Wehrmacht during the Second World War.

Gewehr 43 Rifle

Calibre: 7.92mm
Operation: gas, semi-automatic
Feed: 10-round box magazine
Weight: 9.7lb (4.4kg)
Length: 44in (1117mm)
Sights: mount fitted as standard, often using a 4-power *Zundblickfernrohr* telescopic sight, plus iron sights
Muzzle velocity: 2546fps (776m/s)

Developed in response to the Tokarev rifle used by the Soviet Red Army on the Eastern Front, the Gewehr 43 was a semi-automatic rifle designed both for conventional and sniper use. As a sniper rifle it was not a great success, most top marksmen preferring the simpler and more accurate 98k.

Heckler and Koch G3 SG/1 Sniper Rifle

Calibre: 7.62×51mm NATO
Operation: delayed blowback, selective fire
Feed: 20-round box magazine
Weight: 12.22lb (5.54kg) empty with telescope
Length: 40.4in (1025mm)
Sights: Zeiss or Schmidt and Bender 1.5–6 variable-power telescopic sight, plus iron sights
Muzzle velocity: 2560–2625fps (780–800m/s)

This sniper rifle version of the famous G3 rifle, the SG/1, is used by German police and special forces units, and by other police forces in Europe. Reasonably accurate to 600 metres it is not a precision weapon but it has the advantage of enabling a marksman to fire on full automatic as well as single-shot, and consequently makes a suitable observer's rifle.

Heckler and Koch PSG–1 Sniper Rifle

Calibre: 7.62×51mm NATO
Operation: delayed blowback, semi-automatic
Feed: 5- or 20-round box magazine
Weight: 17.8lb (8.1kg) empty with telescope
Length: 47.6in (1208mm)
Sights: 6×42 Hensoldt Wetzlar telescopic sight

The most highly regarded of semi-automatic sniper rifles, the PSG–1 lives up to its manufacturer's designation as the *Präzisionsschützengewehr* (high-precision marksman's rifle) PSG–1. Along with other Heckler and Koch rifles, the PSG–1 employs the delayed roller-locking system, but also incorporates other features, including a near silent loading and locking action, adjustable trigger weight and width, a stock fully adjustable for height and length, and a bipod and lightweight tripod. Its high cost and weight precludes its use in conventional military sniping, but it is included within the armouries of special forces and police forces around the world.

Heckler and Koch MSG90 Sniper Rifle

Calibre: 7.62×51mm NATO
Operation: delayed blowback, semi-automatic
Feed: 5- or 20-round box magazine
Weight: 14.11lb (6.4kg) empty
Length: 45.9in (1165mm)
Sights: 12-power telescopic sight

A simplified version of the PSG–1, the MSG90 was introduced in 1987 to meet military specifications. The butt stock is adjustable in length and height and a bipod is fitted to the fore-end.

Heckler and Koch BASR Sniper Rifle

Calibre: multiple, including 7.62×51mm NATO, .30–06, .300 Winchester Magnum
Operation: bolt action

Feed: single shot; 4-round integral magazine (3-round in Magnum calibre); or, following modification, 20-round detachable box magazine
Weight: 10lb (4.54kg) heavy-barrelled version, empty without telescope
Length: 42in (1067mm)
Sights: optical sight to be fitted by customer (no iron sights)

A departure from H & K's production of semi-automatic weapons, the BASR (Bolt–Action Sniper Rifle) is intended to fill the gap in the bolt-action market. Designed in a variety of calibres and barrel weights, the BASR is a reasonably priced coverall rifle. The Kevlar stock is manufactured in a classic sporting/hunting configuration.

Mauser SP66 Sniping Rifle

Calibre: 7.62×51mm NATO
Operation: bolt action
Feed: 3-round integral magazine
Sights: neither optical nor iron sights are fitted

Intended both for military and police use the SP66 is a well-made weapon, resembling a target rifle with its finely contoured wooden stock, thumb hole and adjustable cheek rest. The rifle utilizes a short-action Mauser bolt and a very fast lock time, which ensures a speedy exit of the bullet from the muzzle. Although no sights are fitted, Mauser recommend a Zeiss Diavari 1.5–6 variable-power telescopic sight.

Mauser 86SR Sniper Rifle

Calibre: 7.62×51mm NATO
Operation: bolt action
Feed: 9-round double-row box magazine
Weight: 10.8lb (4.9kg) without sight
Length: 47.7in (1210mm)
Sights: mount attached to take a telescopic sight

A development from the SP66, the 86SR features a new bolt action and a ventilated wood-pattern stock to aid heat dissipation from the barrel.

ISRAEL

Galil Sniper Rifle

Calibre: 7.62×51mm
Operation: gas, semi-automatic
Feed: 20- or 25-round detachable box magazine
Weight: 14.11lb (6.4kg) empty; 18.3lb (8.3kg) loaded with telescope
Length: 43.9in (1115mm) with butt extended; 33.1in (840mm) with butt folded
Sights: 6×40 Nimrod telescopic sight, plus iron sights
Muzzle velocity: 2560fps (780m/s)

A rechambered and upgraded version of the Galil assault rifle, it is fitted with a heavy barrel with a muzzle brake and compensator. A bipod is attached to the gas block alongside the forward hand guards. The wooden butt stock has an adjustable cheek piece and rubber recoil pad, and the stock can be folded to reduce overall length when the rifle is being carried. Rather than attempt to mount the telescopic sight directly on to the sheet-metal receiver (with attendant loss of accurate sighting), it is attached to a side rail, although this means the sight is offset-mounted to the left.

Sirkis M36 Sniper Rifle

Calibre: 7.62×51mm
Operation: gas, semi-automatic
Feed: 20-round box magazine
Weight: 9.92lb (4.5kg) without telescope
Length: 33.5in (850mm)
Sights: mount for telescopic sight, plus iron sights

An innovative bullpup format makes the M36 a highly compact rifle, although, perhaps surprisingly, its action is a modification of the US M14. The barrel is full-length and fitted with a combined flash hider and muzzle brake. The in-line stock is manufactured from synthetic materials, to which is attached a bipod and a rubber recoil pad.

ITALY

Beretta Sniper Rifle

Calibre: 7.62×51mm NATO
Operation: bolt action
Feed: 5-round detachable box
Weight: 9.92lb (4.5kg) empty without telescope
Length: 45.9in (1165mm)
Sights: 1.5–6×42 Zeiss Diavari adjustable-power telescopic sight

The sniper rifle of the Italian Army, the Beretta has a wooden stock with thumb hole, adjustable cheek piece and rubber recoil pad. A bipod can be fitted to an attachment at the fore-end of the stock. The rifle's heavy barrel is fitted with a flash-hider.

JAPAN

Sniper's Rifle Type 97

Calibre: 6.5mm
Operation: bolt action
Feed: 5-round box magazine
Weight: 9.25lb (4.2kg)
Length: 50.2in (1275mm)
Sights: 2.5-power telescopic sight, plus iron sights
Muzzle velocity: 2700fps (823m/s)

The sniper version of the standard Type 38 (1905) service rifle, the Type 97 was modified to include a revised bolt and bolt cover, provision for a telescopic sight and folding monopod attached to the front band on the stock. During the Second World War, a new calibre of 7.7mm (.303in) was adopted, and the revised rifle was known as the Type 99 (1939). A sniper version had a 4-power telescopic sight attached.

NORWAY

NM149S Sniper Rifle

Calibre: 7.62×51mm NATO
Operation: bolt action
Feed: 5-round box magazine
Weight: 12.35lb (5.6kg) with telescope
Length: 44.1in (1120mm)
Sights: 6×42 Schmidt and Bender telescopic sight

Employing the well-tried Mauser M98 action, the NM149S is in service with the Norwegian Army and police force. A conventional sniper rifle, it has a laminated beech veneer stock which is adjustable in length through the use of butt spacers. A bipod can be fitted if required.

SOVIET UNION

Mosin-Nagant M1891/30 Rifle

Calibre: 7.62×54mm Rimmed
Operation: bolt action
Feed: 5-round integral box magazine
Weight: 8.82lb (4kg) empty
Length: 48.5in (1232mm)
Sights: either 4-power PE or the lighter 3.5-power PU telescopic sight, plus iron sights
Muzzle velocity: 2660fps (811m/s)

The Mosin–Nagant rifle used by the Red Army during the Second World War was the result of the 1930 modernization of the 1891 pattern rifle. A basic weapon, improved models were selected as sniper rifles, with PE or PU telescopic sights, offset mounted on the left of the receiver.

Tokarev SVT40 Rifle

Calibre: 7.62×54mm Rimmed
Operation: gas, semi-automatic

Feed: 10-round detachable box magazine
Weight: 8.58lb (3.89kg)
Length: 48.1in (1222mm)
Sights: 3.5-power PU telescopic sight, plus iron sights
Muzzle velocity: 2723fps (830m/s)

The SVT (*Samozariadnyia Vintovka Tokoreva* – Tokarev Self-loading Rifle) 40 was an improved version of the highly innovative SVT38 semi-automatic rifle. The SVT40 was developed to give increased firepower to the infantry section, and a few models with telescopic sights were issued to snipers. Ammunition feed jams and problems with the gas system in cold weather, however, led to its removal from front-line service from 1942 onwards.

Dragunov SVD Sniper Rifle

Calibre: 7.62×54mm Rimmed
Operation: gas, semi-automatic
Feed: 10-round box magazine
Weight: 9.5lb (4.31kg) empty with telescope
Length: 48.2in (1225mm)
Sights: 4×24 PSO-1 telescopic sight, with IR detector, plus iron sights
Muzzle velocity: 2723fps (830m/s)

The SVD (*Snayperskaya Vintovka Dragunova* – Dragunov Sniping Rifle) was introduced in the late 1950s as a purpose-built sniper rifle to replace the M91/30 rifle. The pressed-steel receiver and the bolt action (albeit a short-stroke piston system) are similar to that of the Kalashnikov series. Distinctive features include an open 'anatomical' stock and the PSO-1 sight, which has good light-collecting properties and an integral battery for reticle illumination in poor light, and an infra-red detector.

SWITZERLAND

SIG–Sauer SSG 2000 Sniper Rifle

Calibre: 7.62×51mm NATO
Operation: bolt action
Feed: 4-round box magazine

Weight: 14.55lb (6.6kg) with telescope
Length: 47.7in (1210mm)
Sights: 1.5–6×42 variable-power Schmidt and Bender or Zeiss Diatal
    ZA 8×56 telescopic sight
Muzzle velocity: 2460fps (750m/s)

The SSG 2000 utilizes the Sauer 80/90 bolt action, providing a swift and easy loading action. The barrel is fitted with a combined flash suppressor and muzzle brake, and the stock is adjustable with a thumb hole to provide a better grip for the firer's right hand. A bipod or lightweight tripod can be attached to the fore-end. As well as the standard 7.62mm NATO round, the rifle can be chambered for .300 Weatherby Magnum, and Swiss 7.5×51mm cartridges.

SIG SSG 550 Sniper Rifle

Calibre: 5.56×45mm
Operation: gas, semi-automatic
Feed: 20- or 30-round box magazine
Weight: 15.44lb (7.1kg) empty
Length: 44.5in (1130mm); 35.7in (905mm) with butt folded
Sights: mount for telescopic and night sights

The SSG 550 is one of the few sniper rifles manufactured in 5.56mm calibre, developed from the SG 550 assault rifle. Accuracy is improved by the addition of a heavy barrel and a sensitive double-pull trigger. The butt-stock is fully adjustable and can be folded away for carriage. A bipod is fitted as standard.

UNITED KINGDOM and COMMONWEALTH

Whitworth Rifle

Calibre: .45in (11.43mm ball)
Operation: muzzle-loading single-shot
Weight: 8.95lb (4.01kg) without telescope
Length: 49in (1245mm)
Sights: side-mounted Davidson telescopic sight, plus iron sights

Developed during the 1850s, the Whitworth was a military rifle that had the accuracy of a target rifle. A prime reason for the rifle's superior accuracy lay in its precision-made hexagonal bore which fired a bullet moulded to fit the bore. The rifle was almost fully stocked with a good quality wood, normally oak or walnut, with robust iron fittings.

Ross Rifle, Mk III

Calibre: .303in (7.7 × 56mm Rimmed)
Operation: bolt action
Feed: 5-round box
Weight: 9.88lb (4.48kg) without telescope
Length: 50.6in (1285mm)
Sights: various telescopic sights, including 6-power Warner & Swasey Mod. 1913, plus iron sights
Muzzle velocity: 2600fps (792m/s)

The Canadian Ross rifle was adopted by the Canadian Army in 1905, and saw action in the earlier stages of the First World War. Unsuited to the conditions of trench warfare it was subsequently withdrawn, but the Ross was an accurate rifle and numbers were retained for sniping purposes.

Rifle No. 1, Mk III

Calibre: .303in (7.7 × 56mm R)
Operation: bolt action
Feed: 10-round box magazine
Weight: 8.66lb (3.93kg) without telescope
Length: 44.5in (1132mm)
Sights: various telescopic sights, including 2-power Periscopic Prism, 5-power Winchester, 3-power Aldis and 3-power Evans, plus iron sights
Muzzle velocity: 2080fps (634m/s)

The standard rifle of the British Army during the First World War, the No. 1 Mk III was not particularly accurate but large numbers were used in trench sniping. Popularly known as the SMLE (Short Magazine Lee-Enfield), it was an excellent service rifle, ideal for combat conditions and capable of a high rate of fire.

Rifle No. 3, Mk I (Pattern 1914)

Calibre: .303in (7.7 × 56mm R)
Operation: bolt action
Feed: 5-round box magazine
Weight: 9.6lb (4.35kg) without telescope
Length: 46.3in (1175mm)
Sights: various telescopic sights, including the 3-power Pattern 1918 and 3-power Aldis, plus iron sights
Muzzle velocity: 2500fps (762m/s)

The introduction of the SMLE caused concern among the Bisley School of rifle design, and the No. 3 Mk I was developed as a back-up rifle. Utilizing the Mauser bolt-locking system, this was a considerably more accurate rifle than the SMLE, and although awkward to use it was a good sniping weapon. Numbers of this rifle were shipped to America where it was known as the M1917 or 'Enfield rifle'. The No. 3 Mk I was still in use during the Second World War.

Rifle No. 4, Mk I(T)

Calibre: .303in (7.7 × 56mm R)
Operation: bolt action
Feed: 10-round box magazine
Weight: 9.13lb (4.14kg) without telescope; 11.63lb (5.28kg) with telescope
Length: 44.5in (1130mm)
Sights: 3-power No. 32 telescopic sight, plus iron sights
Muzzle velocity: 2465fps (751m/s)

Introduced in 1942 this rifle was the sniper version of the No. 4 Mk 1, which had replaced the SMLE as the British Army's service rifle. The sniper versions were superior examples selected from standard production batches, and then rebuilt and restocked before issue to snipers. A wooden cheek rest was screwed into the comb of the stock to allow the sniper a suitable stock weld when using the Mk 32 telescopic sight.

L42A1 Sniper Rifle

Calibre: 7.62×51mm NATO
Operation: bolt action
Feed:10-round box magazine
Weight: 9.77lb (4.43kg) without telescope
Length: 46.5in (1181mm)
Sights: 3-power L1A1 (redesignated Mk 32) telescopic sight, plus iron sights
Muzzle velocity: 2750fps (838m/s)

The L42A1 was a modification of the No. 4 Mk 1(T), rechambered to take the 7.62mm cartridge. The most obvious difference between the two rifles was the cutting back of the L42A1's fore-end, so that the barrel protruded 15 inches. The L42A1 was only a stop-gap measure, but one that saw service in the British Army until the mid-1980s, by which time it was something of an antique.

Model 82 Parker Hale Sniper Rifle

Calibre: 7.62×51mm NATO
Operation: bolt action
Feed: 4-round integral magazine
Weight: 10.58lb (4.8kg) empty without telescope
Length: 45.8in (1162mm)
Sights: various telescopic sights, according to customer preference, plus iron target sights

The M82 is a straightforward sniper/target rifle utilizing a Mauser bolt action and a heavy free-floating barrel. The wood stock can be extended using butt spacers. This model was taken up by the Canadian Army in the 1970s and modified as the C3 with a 6-power scope, and subsequently by the Australian Army as the L3.

Model 85 Parker Hale Sniper Rifle

Calibre: 7.62×51mm NATO
Operation: bolt action
Feed: 10-round detachable box magazine
Weight: 12.57lb (5.7kg) with telescope

Length: 45.3in–47.7in (1150–1210mm)
Sights: 6×42 Schmidt and Bender telescopic sight, plus emergency iron sights

The Model 85 is a first-rate sniper rifle, capable of precision fire to ranges of 900 metres. The synthetic stock is manufactured in different camouflage colours, according to preference, and is fitted with butt spacers to adjust overall length. A detachable bipod is fitted as standard. The M85 has been taken into British Army service to supplement the L96A1.

Accuracy International PM Sniper Rifle System

L96A1 (PM Infantry)

Calibre: 7.62×51mm NATO
Operation: bolt action
Feed: 10- or 12-round detachable box magazine
Weight: 14.33lb (6.5kg) empty without telescope
Length: 44.3in–47in (1124–1194mm)
Sights: PM 6×42 Schmidt and Bender telescopic sight, plus iron sights

The L96A1 is the British Army designation of the standard PM version of this very accurate and versatile sniper rifle system. The PM utilizes an aluminium frame over which is placed a high-impact plastic stock, incorporating a thumb hole and butt spacers. An adjustable bipod is fitted as standard. An uprated version has been manufactured – the AW – which features many minor improvements, including an easier bolt action, frost-proof mechanism, muzzle brake and a 10×42 Hensoldt telescopic sight with Beta light reticle illumination and an extended hood.

PM Counter-Terrorist Sniper Rifle

Calibre: 7.62×51mm NATO
Operation: bolt action
Feed: 10-round detachable box magazine
Weight: 14.33lb (6.5kg) empty without telescope
Length: 44.3in–47in (1124–1194mm)
Sights: PM 12×42 Military telescopic sight

A modified version of the PM Infantry, this rifle features a more powerful sight and a spring-loaded monopod fitted to the butt, enabling the sniper to keep his weapon on target for long periods of time. Other modifications include a spiral flash hider and compensator. Like others in the series this rifle has been uprated and redesignated as the AWP, a sniper rifle intended for special forces and police operations. Among several improvements is the inclusion of a 3–12×50 variable-power telescopic sight.

PM Covert Sniper Rifle

Calibre: 7.62×51mm NATO subsonic
Operation: bolt action
Feed: 10-round detachable box magazine
Weight: 14.33lb (6.5kg) empty without telescope
Length: 49.39in (1250mm)
Sights: PM 6×42, 10×42 or 12×42 Schmidt and Bender telescopic sights
Muzzle velocity: 1030–1083fps (314–330m/s)

A suppressed version of the PM, this rifle uses special subsonic ammunition and is capable of accurate shooting to ranges as far as 300 metres. Very much a specialist weapon, it is produced in a 'take-down' format so that it can be dismantled to fit into an airline suitcase. Following the introduction of the AW series, this rifle has now been redesignated the AWS.

PM Super Magnum Sniper Rifle

Calibre: .338in (8.6×70mm) Lapua Magnum
Operation: bolt action
Feed: 5-round box magazine
Weight: 14.99lb (6.8kg) empty without telescope
Length: 50in (1268mm)
Sights: 10×42 Bausch & Lomb telescopic sights
Muzzle velocity: 3000fps (914m/s)

A prime example of the medium-heavy sniping rifle, this weapon can also be ordered in .300 Winchester and 7mm Remington

Magnum calibres. The rifle is designed for anti-personnel sniping at distances over 1000 metres, as well as for the destruction of light armour and other military equipment at long ranges. The rifle has subsequently been redesignated the SM (Super Magnum) sniper rifle.

UNITED STATES

M1903A1 Springfield Rifle

Calibre: .30–06in (7.62 × 63mm)
Operation: Bolt action
Feed: 5-round box magazine
Weight: 9lb (4.1kg)
Length: 43.5in (1105mm)
Sights: various telescopic sights, plus iron sights
Muzzle velocity: 2805fps (855m/s)

Introduced as the standard infantry weapon of the US armed forces in the early years of the century, the M1903 Springfield was a reliable and accurate rifle, which saw service in both world wars and Korea. The M1903A1 was an inter-war modification and featured a redesigned stock with a form of pistol grip; this rifle was used by the US Marines in the Pacific, fitted with an 8-power Unertl telescopic sight. The M1903A4 was a specially produced sniper variant with a revised bolt and 3-power Weaver telescopic sight, but no iron sights. This variant saw service with the US Army in Europe during the Second World War and later in Korea.

M1C/D Garand Sniper Rifle

Calibre: .30–06in (7.62 × 63mm)
Operation: gas, semi-automatic
Feed: 8-round clips
Weight: 9.5lb (4.32kg)
Length: 43.6in (1107mm)
Sights: Griffen and Howe or 2.5-power M81 telescopic sights (M1C), and M81 or 2.5-power M82 telescopic sights (M1D), plus iron sights.
Muzzle velocity: 2805fps (855m/s)

The US army issued two sniper models of the Garand rifle, varying only in the differing mounts for the telescopic sight. A good service rifle the Garand was somewhat awkward for snipers, particularly in the clip feed which necessitated offset sights and a leather cheek pad to allow a proper stock weld. A short, funnel-shaped flash-hider was fitted over the muzzle.

Winchester Model 70

Calibre: .30–06in (7.62×63mm)
Operation: bolt action
Feed: 5-round integral magazine
Sights: 8-power Unertl telescopic sight

The Winchester M70 was a sporting rifle taken up by the US Marines for recreational and target shooting purposes. Marine snipers saw the possibilities of this accurate weapon and sight combination, and used it in combat in the Korean and Vietnam wars.

M40 Sniper Rifle

Calibre: 7.62×51mm NATO
Operation: bolt action
Feed: 5-round magazine
Weight: 6.84lb (3.1kg) empty without telescope
Length: 41.6in (1055mm)
Sights: 3–9 variable-power Redfield AccuRange telescopic sight
Muzzle velocity: 2790fps (850m/s)

The military version of the Remington Model 700 sporting rifle, it was used by the US Marines in Vietnam. Employing the Mauser bolt-action the M40 was an accurate weapon, capable of long-range shooting, and a superior sniper rifle to military models during the 1960–70s.

M40A1 Sniper Rifle

Calibre: 7.62×51mm NATO
Operation: bolt action

Feed: 5-round magazine
Weight: 14.5lb (6.57kg) without telescope
Length: 44in (1117mm)
Sights: 10-power Unertl telescopic sight
Muzzle velocity: 2550fps (777m/s)

Aware of the limitations of the essentially civilian M40, the US Marines tested a variety of bolt-action and semi-automatic weapons before deciding upon a conventional bolt-action rifle. Utilizing the same M700 Remington receiver as the M40, the M40A1 has a heavy 26-inch stainless steel barrel and a fibreglass stock with epoxy filler. Fitted with a powerful telescopic sight the M40A1 is accurate to a range of 1000 yards.

M21 Sniper Rifle

Calibre: 7.62×51mm NATO
Operation: gas, semi-automatic
Feed: 20-round detachable box magazine
Weight: 11.25lb (5.1kg) empty without telescope
Length: 44.1in (1120mm)
Sights: 3–9 variable-power Redfield AccuRange telescopic sight, plus iron sights
Muzzle velocity: 2805fps (855m/s)

The M21 is the redesignation of the M14 National Match/Accurized rifle, an upgraded version of the standard M14. The main improvements comprise a specially chosen walnut stock impregnated with an epoxy resin and glass fibre bedded with the receiver, modified firing mechanism and a higher specification barrel. The M21 fires only on semi-automatic and not full automatic. An improved version is being manufactured commercially by the Springfield Armory, with a heavier barrel, adjustable cheek piece, recoil pad, bipod and improved telescopic sight.

M24 Sniper Weapon System

Calibre: 7.62×51mm NATO
Operation: bolt action
Feed: 5-round integral magazine

Weight: 12.1lb (5.49kg) empty without telescope
Length: 43in (1092mm)
Sights: 10×42 Leupold Ultra M3 telescopic sight, plus detachable emergency iron sights

The M24 represents a return to bolt-action sniper rifles by the US Army. As in the US Marine M40A1, the M24 uses the Remington 700 action, although the receiver has been made for adaptation to take the .300 Winchester Magnum round. The stock is made of a composite of Kevlar, graphite and fibreglass bound together with epoxy resins, and features an aluminium bedding block and adjustable butt plate. A detachable bipod can be attached to the stock's fore-end.

RAI Model 300 Convertible Long-Range Rifle

Calibre: 7.62×51mm NATO, or 8.58×71mm RAI
Operation: bolt action
Feed: 5-round box magazine (7.62mm), 4-round magazine (8.58mm)
Weight: 12.7lb (5.67kg) empty without telescope
Sights: mount for telescopic sight, according to customer preference
Muzzle velocity: 7.62mm – 2625fps (800m/s); 8.58mm – 3000fps (914m/s)

As its name suggests, this rifle has been designed for two calibres from the outset; by changing the barrel and bolt-head the 7.62mm rifle can be transformed to accept the 8.58mm cartridge developed by RAI (similar to the .338 Lapua Magnum round). Unorthodox in appearance the Model 300 has a fluted heavy barrel, which stands clear of the fore-end, and an open 'anatomical' butt stock which can be adjusted for length. A bipod is fitted to the fore-end, which also contains a harmonic balancer to dampen vibration.

H-S Precision HSP762/300 Sniper Rifle

Calibre: 7.62×51mm NATO, or .300 Winchester Magnum (7.62×66mm)
Operation: bolt action
Feed: 4-round box magazine (7.62mm); or 3-round box magazine (.300in)
Weight: 12.5lb (5.67kg) without scope

Sights: 10-power Bausch & Lomb Mk 1, or 10-power Leupold Ultra M1/M3

Muzzle velocity: 7.62mm: 2560fps (780m/s); .300in: 2840fps (865m/s)

Not only does this rifle come in two calibres (with two barrels and bolts) but it has a full take-down capability, which enables the rifle to be packed away in a 23×17in case. The Kevlar/graphite/fibreglass ¬tock comes with an adjustable butt plate (similar to that of the M24) and a bipod. The fluted barrel is locked into the receiver by an interrupted screw thread and a matching bracket plate. Unlike many take-down rifles, the HSP762/300 is highly accurate and acts as a versatile sniper rifle.

SR–25 Sniper Rifle

Calibre: 7.62×51mm NATO
Operation: gas, semi-automatic
Feed: 20-round detachable box magazine
Weight: 10.75lb (4.88kg) empty without telescope
Length: 44in (1117mm)
Sights: mount attached for a variety of telescopic sights

Developed by Eugene Stoner, this rifle exhibits some of the characteristics of his former designs, notably the AR–10 and the M16. However, this is not an uprated assault rifle but a sniper's weapon, of special use to those who require a facility for rapid follow-up shots.

Barrett Model 82A1 Sniper Rifle

Calibre: .5in (12.7×99mm) Browning
Operation: short recoil, semi-automatic
Feed: 11-round detachable box magazine
Weight: 29.5lb (13.4kg) without telescope
Length: 61in (1550mm)
Sights: 10-power Leupold and Stevens M3a Ultra telescopic sight
Muzzle velocity: 2800fps (853m/s)

Combat proven in the Gulf War, the Barrett 'Light Fifty' is the only semi-automatic .50-calibre sniper rifle on the market. When the bullet is fired, the barrel is sent backwards 2 inches, absorbing recoil, a function also carried out by a large muzzle brake. The rifle comes with a bipod and can be mounted on to a machine-gun tripod if required. Barrett have also produced a bullpup version, called the M90, which is 16 inches shorter than the M82A1.

McMillan M1987R

Calibre: .5in (12.7×99mm) Browning
Operation: bolt action
Feed: 5-round box magazine
Weight: 25lb (11.34kg) with telescope
Sights: 10-power or 16-power telescopic sights of various manufacture

One of the lighter of the .50-calibre sniper weapons, the M1987R is a conventionally configured bolt-action rifle. The extensive recoil is absorbed by a 'pepperpot' muzzle brake and recoil pads. A bipod is attached to the synthetic stock. McMillan have also produced a variant, the 'Combo 50', for the US Navy SEALs, which features a folding stock.

RAI Model 500 Long-Range Rifle

Calibre: .5in (12.7×99mm) Browning
Operation: bolt action
Feed: single shot
Weight: 30lb (13.6kg) without telescope
Sights: mount for telescopic sights, according to customer preference.
Muzzle velocity: 2914fps (888m/s)

The layout of this weapon is similar to the RAI Model 300, with a long fluted barrel and an open butt stock, adjustable for length and height. The fore-end contains a harmonic balancer which damps out vibrations from the barrel. The action is unusual, in that the very short bolt is removed completely during the loading manoeuvre, and replaced as the cartridge is pushed into the chamber. The rifle is in service with police forces, the US Navy and the US Marines.

YUGOSLAVIA
M76 Sniper Rifle

Calibre: 7.92×57mm
Operation: gas, semi-automatic
Feed: 10-round detachable box magazine
Weight: 9.26lb (4.2kg) without telescope
Length: 45.4in (1153mm)
Sights: 4-power ON M76 telescopic sight
Muzzle velocity: 2362fps (720m/s)

A derivative of the Soviet SVD, the M76 adopts the old German 7.92mm calibre used in Yugoslavia before the Second World War (although, however, variants are manufactured to take the Soviet 7.62×54mm and NATO 7.62×51mm cartridges). In contrast to the skeletonized butt stock of the SVD, the M76 is fitted out with a wooden stock, pistol grip and hand guard. The telescopic sight is similar to the Soviet PSO-1, but does not use a battery operated light to illuminate the reticle; instead it employs tritium illumination, which casts a green glow into the sight when light levels are low.

CHAPTER SIXTEEN

AMMUNITION

The importance accorded to rifles as the determining factor of success in sniping can lead to a displacement of the key role of ammunition. If ammunition is substandard then the best rifle in the world will perform badly. For the sniper, ammunition must be of superior quality so that it will perform consistently in all conditions. Accuracy is dependent on the consistency of the round, so that each time the trigger is pulled the bullet will travel up the barrel at the same velocity as previous bullets, and will describe the same ballistic trajectory. If this consistency of operation is not forthcoming then the ammunition will be unsuitable for sniping purposes. As a result, snipers use match-grade ammunition which is manufactured under stringent conditions of quality control.

Small-arms ammunition consists of a cartridge case, a primer, a propellant charge and a bullet.[1] The modern rifle round was developed in the latter part of the nineteenth century and still remains the basis for sniper ammunition.

The cartridge case acts as the container for the primer, propellant and bullet. It must be sufficiently strong to withstand the rigours of handling in combat conditions, be resistant to corrosion and prevent the ingress of moisture. In addition, the case must be flexible enough to expand to seal the chamber when the bullet is fired (obturation) and be able to return to its original size to enable easy extraction. The material most able to meet these requirements is brass (70 per cent copper, 30 per cent zinc) which combines strength and elasticity with a good resistance to corrosion.

In describing the calibre of a military round (using metric measurement), it is customary to refer to both the diameter of the

261

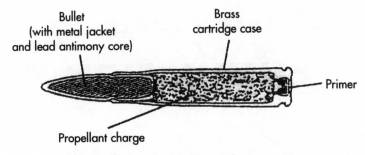

Figure 7: Standard 7.62mm Rifle Ammunition.

rifle bore and the length of the cartridge case, as a way of providing an indication of ammunition performance. Thus, for example, the standard 7.62×51mm NATO round has a bullet diameter of 7.82mm, designed to fit a bore of 7.62mm (the bullet is always slightly bigger than the bore, to force it into the rifling), and a cartridge case length of 51mm.

Although something of a misnomer, most cartridge cases are termed rimless, which despite having a rim (to aid extraction) actually means that the rim is formed by the recessed groove within the wall of the case and does not protrude outwards. Protruding rimmed rounds were found to be difficult to operate in automatic-firing magazines, although some have had a long life, and include the British .303in and the Russian/Soviet 7.62× 54mm R (hence the 'R' suffix).

The primer is held in a recessed pocket in the base of the cartridge case. Once struck by the firing pin, the composition in the primer is ignited, the flash travelling along a fire hole (or holes) to the propellant in the body of the case. Primers were made from fulminate of mercury or potassium chlorate, or a combination of both. Although providing a good flash, these materials led to case corrosion and rusting in the rifle barrel, and have been replaced by lead styphenate, which does not have these deleterious effects yet provides suitable ignition.

The propellant provides the motive force which sends the bullet hurtling up and out of the barrel. The power of a few grains of propellant is enormous. A fairly standard round (the British .303in Mk VII) develops a pressure of 18 tons per square inch at the point of ignition; this spins the bullet at a rate 2640 revolutions per *second* as it leaves the muzzle at a velocity of 2500 feet per second (Rifle No. 3, Mk I). The bullet has a maximum range of 3400 yards, and when fired in the vertical plane it will rise to a maximum height of 9200 feet, taking 17 seconds to reach its apogee and taking 45 seconds to return to the ground, still spinning and base first.[2] And when a full-power, high-velocity bullet hits the human body, at ranges of anything up to a mile, the results are suitably devastating.

The modern rifle round owes much to the development of celluloid in the 1870s. Nitric acid was added to celluloid fibres, and then mixed with alcohol and ether to form a gelatinous medium which then could be rolled into thin sheets, dried and cut into particles of nitrocellulose. The use of celluloid allowed the burn rate of the propellant to be regulated, in contrast to gun cotton (cotton soaked in nitric acid) which was unsuitable for small arms. Although called powder (from gunpowder days), nitrocellulose is, in fact, manufactured into one of three granular shapes – ball, flake and stick – according to the production process used. The size of the granules varies the burning speed of the propellant; a slower burn rate is generally preferred in rifle ammunition.

In the late nineteenth century, the single-base propellant (nitrocellulose) was supplemented by a double-base propellant (nitrocellulose and nitroglycerine). Nitroglycerine increased the energy of the powder although it was subsequently discovered that its greater flame temperature reduced the life of the barrel. As a consequence, nitroglycerine is used only in small quantities in modern ammunition. Some propellants – such as that used in the NATO 7.62mm round – remain single-based and contain approximately 98 per cent nitrocellulose, the remainder comprising preservatives and stabilizers to improve shelf-life, lubricants to reduce barrel wear and additives to cut down muzzle flash.[3]

The bullet is the last component of small-arms ammunition, and is the active element which produces the end result. Musket balls were made of lead, and even today lead antimony (a slightly harder compound) is used to make rifle bullets (the standard anti-personnel bullet is commonly referred to as a 'ball' from this historical association). Lead antimony has the advantages of being sufficiently heavy to carry at long ranges and hit the target with sufficient power, and yet be soft enough to deform slightly at the moment of ignition and grip the rifling in the barrel.

The introduction of the high-velocity propellant nitrocellulose placed too great a stress on the lead bullet, so it was enclosed in an envelope of a harder metal, usually either cupro-nickel or a more economical gilding metal of copper and zinc. The result was known as a full metal jacket (FMJ) bullet, although the base of the bullet was left unjacketed to allow the force the ignition to push the soft lead core into the rifling.

The shape of the rifle bullet is governed by its requirement to maintain high-velocity at long ranges. As a consequence it is designed to be narrow with a long-pointed tip so that it cuts through the air as efficiently as possible. To reduce base drag, bullets intended for sniper use are usually made in a boat-tail configuration, that is, the bullet is tapered at the base, reminiscent of the stern of a boat. As with all other components of the round, the bullet must be manufactured to the highest standards to maintain a consistent level of performance.

Since the 1970s, work has been ongoing in transforming the rifle ammunition from its standard configuration. An experimental round has been developed that dispenses with the traditional case: the propellant itself is moulded into the right shape so that it will fit directly into the rifle chamber. Other unorthodox designs concentrate on replacing the conventional bullet with multiple projectiles, fleschettes and discarding sabot rounds. At present they remain in the experimental stage and are unlikely to replace brass cartridge ammunition for some time. This is especially true of the sniper, who values consistency of operation over other factors, such as an economical and lightweight round. The development of sniper ammunition will take place through the introduction of a more powerful cartridge to replace the standard 7.62×51mm NATO round. As many such replacements already exist, progress in this field is more a question of logistics than technology.

NOTES

Only those works not listed in the Bibliography
have been given full publication details here.

CHAPTER ONE
RIFLEMEN AND SKIRMISHERS

1. Peterson, *The Book of the Gun*, p. 138.
2. *The Virginia Gazette*, quoted in Sawyer, *Firearms in American History*, pp. 78–9.
3. Hanger, *To All Sportsmen*, quoted in Moore, *Weapons of the American Revolution*, pp. 61–2.
4. Hanger, op. cit., quoted in Lewis, *Small Arms and Ammunition in the United States Service*, p. 90.
5. Moore, op. cit., p. 64.
6. Peterson, op. cit., pp. 164–5.
7. Baker, *Remarks on the Rifle*, p. 53.
8. Haythornthwaite, *The Napoleonic Source Book*.
9. Blakiston, *Twelve Years Military Adventure*, Vol. II, pp. 344–5, quoted in David Gates, *The British Light Infantry Arm*, (Batsford, 1987) p. 172.
10. Hargreaves, 'The Lonely Art', in *Marine Corps Gazette*, December 1954, pp. 71–2, and Howarth, *Trafalgar* (Collins, 1969), p. 176.
11. Peterson, op. cit., pp. 117–19.
12. Lewis, op. cit., pp. 13–14.

CHAPTER TWO
THE AMERICAN CIVIL WAR

1. Morrow, *The Confederate Whitworth Sharpshooters*, p. 77. Morrow's book is a detailed account of the exploits of these soldiers.
2. Morrow, op. cit., pp. 9, 12, 17; and Lewis, *Small Arms and Ammunition*, p. 104.
3. *Confederate Veterans*, Vol. I, p. 117, quoted in Morrow, op. cit., pp. 29–30.
4. John West, quoted in Morrow, op. cit., p. 43.
5. Morrow, op. cit., p. 43. Morrow also discusses the likely sharpshooter involved, probably Private Charlie Grace of 4th Georgia Regiment.
6. Captain Park, *Southern Historical Society Papers*, Vol. I, p. 380, quoted in Morrow, op. cit., p. 74.

7. Captain Ashe, *Confederate Veterans*, Vol. XXXV, pp. 254–5, quoted in Morrow, op. cit., p. 55.
8. Quoted in Senich, *The Pictorial History of US Sniping*, pp. 1–2.
9. John West, quoted in Morrow, op. cit., p. 42.
10. Major Thomas B. Brooks, *Official Records*, Series 1, Vol. XXVIII, Part I, p. 277, quoted in Morrow, op. cit., p. 56.
11. Morrow, op. cit., p. 62.
12. John W. Green, *Johnny Green of the Orphan Brigade*, pp. 127–8, quoted in Morrow, op. cit., p. 40.
13. William C. Davis, *The Orphan Brigade*, pp. 215–16, quoted in Morrow, op. cit., pp. 37–8.
14. Davis, op. cit., p. 218, quoted in Morrow, op. cit., p. 39.
15. *Confederate Veterans*, Vol. XVI, p. 172, quoted in Morrow, op. cit., p. 79.
16. 'T.E.C.', *Civil War Times Illustrated*, pp. 135–6, quoted in Morrow, op. cit., pp. 75–6.
17. Charles Stevens, *Berdan's United States Sharpshooters in the Army of the Potomac 1861–65*, quoted in Marcot, *Civil War Chief of Sharpshooters: Hiram Berdan*, p. 120.
18. Stevens, op. cit., quoted in Marcot, op. cit., p. 135.
19. Stevens, op. cit., quoted in Sword, *Sharpshooter: Hiram Berdan, his famous Sharpshooters and their Sharps Rifles*, p. 42.
20. White, unpublished diary, quoted in Marcot, op. cit., p. 122.
21. Sword, op. cit., pp. 40–41.
22. Sword, op. cit., pp. 47–8.
23. Sword, op. cit., p. 61.
24. Thompson, *History of the Orphan Brigade*, p. 243, quoted in Morrow, op. cit., p. 40.
25. Flores, 'Sharpshooters in the Civil War', in *Gun Digest*, 1977, p. 12. quoted in Morrow, op. cit., p. 71.
26. Quoted in Morrow, op. cit., p. 44.

CHAPTER THREE
SNIPING AND NINETEENTH-CENTURY TECHNOLOGY

1. Peterson, *The Book of the Gun*, p. 173.
2. Peterson, op. cit., p. 182.
3. Lumsden (ed), *Anti-Personal Weapons*, p. 79.
4. Lumsden, op. cit., p. 79.
5. Lumsden, op. cit., p. 80.
6. Reitz, *Commando: A Boer Journal of the Boer War*, pp. 39, 42.
7. Shore, *With British Snipers to the Reich*, p. 276.
8. Hargreaves, 'The Lonely Art', in the *Marine Corps Gazette*, December 1954, pp. 73–4.

CHAPTER FOUR
THE SNIPER EMERGES: 1914–16

1. Hesketh-Prichard, *Sniping in France*, p. 28.
2. 'Instructions for the use of S.m.K. Cartridges and Rifles with Telescopic Sights', quoted in Senich, *The German Sniper 1914–1945*, p. 4.
3. Dunn, *The War the Infantry Knew*, p. 83.
4. Tantum, *Sniper Rifles of the Two World Wars*, pp. 17–18.
5. Hesketh-Prichard, op. cit., p. 2.
6. Edmonds (C. E. Carrington), *A Subaltern's War*, pp. 70–71.
7. Crum, *Scouts and Snipers in Trench Warfare*, pp. ix–x.
8. Graves, *Goodbye to All That*, p. 119.
9. *History of the Ministry of Munitions*, Vol. XI, Part III (Crown Copyright), p. 42. So 'helpful' were the Germans that they were even prepared to demobilize special workmen from the Army to enable the orders to be swiftly fulfilled!
10. See Bull, 'British Army Snipers, 1914–18' in *Military Illustrated* for further details of British telescopic sights; his article also provides a clear and informed overview of the subject.
11. McBride, *A Rifleman went to War*, pp. 79–81.
12. Ernest Shephard, *A Sergeant-Major's War: From Hill 60 to the Somme* (Bird/ Crowood, 1987), p. 33.
13. Clarke, *Sniper on the Western Front*, pp. 6–7.
14. Hesketh-Prichard, op. cit., pp. 34–5.
15. Hesketh-Prichard, op. cit., p. 34.
16. See McBride, op. cit., p. 96 and Skennerton, *The British Sniper*, pp. 30–1. See also Idriess, *The Desert Column*, p. 52, where an attempt was made, unsuccessfully, to hit a target at 500 yards range with a sniperscope.
17. Major Shaw, quoted in Robert Rhodes James, *Gallipoli* (Batsford, 1965), p. 158.
18. A. P. Herbert, *The Secret Battle*, pp. 49–50, quoted in Rhodes James, op. cit., pp. 159–60.
19. Idriess, *The Desert Column*, p. 25.
20. McCowan, unpublished ms. diary.
21. Hogue, *Trooper Bluegum at the Dardanelles*, p. 191.
22. Hogue, op. cit., pp. 153–4.
23. Idriess, op. cit., p. 43.
24. Hogue, op. cit., pp. 191–3.
25. Armstrong wrote a useful snipers' manual – *Fieldcraft, Sniping and Intelligence* – during the Second World War in an attempt to promote sniping.
26. Skennerton, op. cit., p. 175.
27. Skennerton, op. cit., p. 219.
28. See Ashworth, *Trench Warfare 1914–18 – The Live and Let Live System* for a detailed study of this phenomenon.

29. For example, Clarke, op. cit., p. 6: '. . . they offered cigs and tea so I would keep moving on'; and Carson Catron, unpublished memoir, p. 2: 'As they moved about their own front line, they [the snipers] would be told, "Push off now, do your stuff somewhere else".'

30. Dunn, op. cit., p. 98.

31. Graves, op. cit., p. 112.

32. Ricketts, unpublished memoir, p. 34.

33. McBride, *A Rifleman Went to War*, pp. 118–19.

34. Ian Hay, *The First Hundred Thousand* (Blackwood, 1916), pp. 210–11.

35. Dunn, op. cit., p. 200.

36. Ricketts, op. cit., p. 33. See also Graves, op. cit., p. 114, where the same point is made.

37. See Dunn, op. cit., p. 84, on killing of sniper found in a haystack, and McCowan, op. cit., 29 April 1915, on the death of a Turkish sniper.

38. Carrington (Edmonds), op. cit., p. 113.

CHAPTER FIVE
A FULLY-FLEDGED ARTICLE: THE SNIPER 1916–18

1. Hesketh-Prichard, *Sniping in France*, pp. 75–88.

2. Hesketh-Prichard, op. cit., p. 12.

3. Armstrong, *Fieldcraft, Sniping and Intelligence*, p. 3.

4. Crum, *Scouts and Snipers in Trench Warfare*, p. 4.

5. Hesketh-Prichard, op. cit., p. 55.

6. Hesketh-Prichard, op. cit., p. 54.

7. Hesketh-Prichard, op. cit., pp. 46–7.

8. Hall, *Recollections, 1915–19*, pp. 4–5.

9. McBride, *A Rifleman Went to War*, pp. 308–10.

10. McBride, op. cit., pp. 317–18.

11. Jünger, *Copse 125*, pp. 117–23.

12. Dunn, *The War the Infantry Knew*, p. 470.

13. Armstrong, op. cit., p. 97.

14. McBride, op. cit., p. 315.

15. See Chesney, *The Art of Camouflage*.

16. Hesketh-Prichard, op. cit., p. 102.

17. Glover, 'Summary of Snipers' Reports', 25/1/1917.

18. Hesketh-Prichard, op. cit., pp. 128–31.

19. Armstrong, op. cit., p. 28.

20. ibid.

21. John W. Thomas jnr, *Fix Bayonets!* (Scribners, 1926), p. 23.

22. Senich, *The Pictorial History of US Sniping*, pp. 36–7.

23. McBride, op. cit., pp. 95–6; and Hesketh-Prichard, op. cit., p. 104.

24. Hesketh-Prichard, op. cit., pp. 103–4.
25. Dunn, op. cit., p. 494.

CHAPTER SIX
THE SECOND WORLD WAR

1. Armstrong, *Fieldcraft, Sniping and Intelligence*, p. v.
2. For example, *Military Training Pamphlet No. 44 – 1940: Notes on the Training of Snipers*, a short yet reasonably comprehensive outline of sniping tactics based largely on experience from the First World War.
3. Shore, *With British Snipers to the Reich*, p. 136
4. Rabbets, IWM audio interview transcript, pp. 36–9.
5. Rabbets, op. cit., p. 58.
6. Rabbets, op. cit., pp. 57–9.
7. Eloise Engle and Lauri Paananen, *The Winter War: The Russo-Finnish Conflict 1939–40* (Sidgwick & Jackson, 1973), p. 136.
8. Tantum, *Sniper Rifles of Two World Wars*, pp. 25–6, and Senich, *The German Sniper*, p. 407.
9. Morozov, 'Sniper Tactics' in *Marine Corps Gazette*, Vol. 27, No. 4.
10. Zaitsev, quoted in Alan Clark, *Barbarossa*, (Hutchinson, 1965). See also Craig Roberts, 'At Stalingrad . . . Sniper versus Sniper', in *World War II*, September 1989, pp. 12–14, 16.
11. Zaitsev, op. cit.
12. Anonymous Soviet source, quoted in Senich, op. cit., pp. 404–5.
13. See Bolotin, *Fifty Years of Soviet Small Arms*, and for a debunking of Soviet claims, Shore, op. cit., pp. 95–7.
14. Senich, op. cit., p. 149.
15. Senich, op. cit., pp. 281–4.
16. Tantum, op. cit., pp. 20–22.
17. Hetzenauer, quoted in Senich, op. cit., p. 115. Captain Hans Widhofner, an Austrian officer, collated the replies to a detailed questionnaire on sniping issued to Hetzenauer and two other sniper colleagues, Sepp Allerberger and Helmut Wirnsberger. The results were published in 1967 in the Austrian journal *Truppendienst*, and reproduced here from Peter Senich's *The German Sniper*, pp. 113–21.
18. 'Sniper: The Invisible Enemy', produced for the OKH (German Army High Command) and reissued by International Historic Films, 1988. See also 'Snipers in Action: The Unseen Weapon', a training film featuring re-enactment studies of sniper actions, produced in 1944 and reissued by International Historic Films.
19. Shore, op. cit., pp. 88–9.
20. Shore, op. cit., pp. 89–90.
21. Quoted in Senich, op. cit., p. 131.
22. 'Snipers in Action', training film, op. cit.

23. Skennerton, *The British Sniper*, pp. 99–103, and Tantum, op. cit., pp. 17–18.
24. Shore, op. cit., pp. 288–91.
25. Jalland, IWM audio interview.
26. Spearman, IWM audio interview transcript, pp. 38–9.
27. Shore, op. cit., p. 323.
28. Shore, op. cit., p. 296.
29. Quoted in Shore, op. cit., p. 110.
30. Shore, op. cit., p. 106.
31. Shore, op. cit., p. 9.
32. Belfield and Essame, *The Battle for Normandy*, pp. 123–4, quoted in Ellis, *The Sharp End of War*, p. 90.
33. Shore, op. cit., pp. 9–10. See also Max Hastings, *Overlord* (Michael Joseph, 1984), pp. 148–9, which quotes a report on the poor morale and performance of a British infantry battalion, one of the alleged factors being the failure of officers and NCOs to wear rank badges.
34. Jalland, op. cit.
35. Shore, op. cit., p. 72.
36. Spearman, op. cit.
37. Tantum, op. cit., p. 14.
38. Senich, *Limited War Sniping*, pp. 26–7.
39. Fulcher, quoted in Sasser and Roberts, *One Shot – One Kill*, pp. 32–3.
40. First US Army Group, 'Battle Experiences', quoted in Doubler, *Busting the Bocage*, p. 27.
41. Hansen diary, US Army Military History Institute, Carlisle, quoted in Hastings, op. cit., pp. 209–10.
42. Tantum, op. cit., p. 31.
43. 'How Japs Train Snipers', in *Military Review*, October 1945, p. 86.
44. ibid.
45. Jalland, op. cit.
46. US Army Center for Lessons Learned, *Bulletin 1–88*, pp. 3–4.
47. Westerfield, *The Jungleers*, p. 3.
48. Westerfield, op. cit., p. 3–4.
49. Shore, op. cit., p. 116.
50. Shore, op. cit., p. 117.
51. Skennerton, op. cit., p. 201.
52. Van Orden and Lloyd, *Equipment for the American Sniper*.
53. Tolbert, 'Deadly Teams Emerge from this Academy', in *Leatherneck*, October 1943.
54. Eric Hammel and John E. Lane, *76 Hours: The Invasion of Tarawa* (Pacifica Press, 1985), pp. 33–4, 47–9, 124, 155–7, 174–5, 235. The commander of the 2nd Marine Regiment's scout-sniper platoon, Lieutenant William Deane Hawkins, was awarded a posthumous Medal of Honor for making repeated attacks on Japanese bunkers, despite being severely wounded.
55. Sasser and Roberts, op. cit., pp. 73–7.

56. Chandler and Chandler, *Death from Afar: Marine Corps Sniping*, p. 28.
57. US Army Ground Forces, Observer Board, Pacific Ocean Areas, Report No. 183: Training and Use of Snipers.

CHAPTER SEVEN
THE KOREAN WAR

1. Krilling, quoted in Senich, *Limited War Sniping*, p. 68.
2. 'Report of Sniping Activities', 25 March 1952, quoted in Senich, op. cit., p. 29.
3. 'Report of Sniping Activities', 25 March 1952, quoted in Senich, op. cit., p. 28.
4. Test Report of Marine Corps Equipment Board, Project 757: Sniper Rifles, Telescopes and Mounts.
5. Hicks, 'Team Shots Can Kill', in *Marine Corps Gazette*, December 1963.
6. Martin, 'They Call Their Shots', in *Marine Corps Gazette*, April 1953.
7. Holmes, quoted in Martin, op. cit.
8. Senich, op. cit., pp. 18, 19–20.
9. Senich, op. cit., p. 20.
10. Quoted in Senich, op. cit., pp. 35–7.

CHAPTER EIGHT
WOUND BALLISTICS

1. Longmore, *Gunshot Injuries*, quoted in Lumsden (ed.), *Anti-Personal Weapons*, p. 54.
2. Owen-Smith, *High Velocity Missile Wounds*, pp. 9–13, and Lumsden, op. cit., pp. 53–7.
3. Lumsden, op. cit., pp. 60–1.
4. Owen-Smith, op. cit., pp. 15–17.
5. Plaster, *The Ultimate Sniper*, pp. 133–6.
6. For a detailed discussion of the mechanisms of injury see Owen-Smith, op. cit., pp. 18–41, Lumsden, op. cit., pp. 62–3, and Kokalis, 'Killing Effect II' in *Soldier of Fortune*, January 1989, pp. 84–7 and 112.
7. Owen-Smith, op. cit., pp. 22–3.
8. From Charles Whiting, *Bloody Aachen*, p. 43, quoted in Ellis, *The Sharp End of War*, p. 113.
9. Plaster, op. cit., pp. 133–4.
10. Kokalis, op. cit., pp. 85–6, and Ed Sanow, quoted in Plaster, op. cit., pp. 130–2.
11. Lumsden, op. cit., p. 93.
12. Kokalis, op. cit., p. 87.
13. Plaster, op. cit., p. 132.
14. Plaster, op. cit., p. 134.

15. Shore, *With British Snipers to the Reich*, p. 324.
16. Shore, op. cit., p. 325.
17. Heaton, quoted in Martin Middlebrook, *Operation Corporate*, p. 335.

CHAPTER NINE
THE WAR IN VIETNAM

1. Childs and Martin, 'Sniper . . .', in *Leatherneck*, January 1967.
2. See, for example: Hofues, 'Modernize the Sniper Rifle', in *Army*, June 1957; Odom, 'The Case of the US Sniper', in *Infantry School Quarterly*, April 1954; and Collins, 'Pro & Con', in *Infantry*, January–March 1959.
3. Hofues, op. cit.
4. See Henderson, *Marine Sniper*, pp. 75–6.
5. Figures reproduced in Sasser and Roberts, *One Shot – One Kill*, p. 132.
6. For details of NVA sniper training, organization and tactics see 'VC/NVA Employment of Snipers' (US MACV Intelligence Center, Vietnam), and Childs, 'VC defector unveils élite sniper company' in *Sea Tiger*, 19 July 1966.
7. Ewell and Hunt, *Sharpening the Combat Edge*, p. 116.
8. Plaster, *The Ultimate Sniper*, p. 400.
9. Ward, *Dear Mom: A Sniper's Vietnam*, p. 151. Ward had been ordered not to engage the sniper because of the impending air strike.
10. 'US Army Center for Lessons Learned', *Bulletin 1–88*, p. 4.
11. Dabney, 'Dabney's Hill', in Page and Pimlott (eds), *NAM: The Vietnam Experience*, p. 333.
12. Philip Caputo, *A Rumour of War*, p. 68.
13. Shulimson and Johnson, *US Marines in Vietnam 1965*, p. 131.
14. 'Sniper Rifle/Telescopic sight, recommendation of' from Scout-Snipers in Current Operations: Routing Sheet A03M–jaj, 26 January 1966.
15. Childs and Martin, op. cit.
16. Childs and Martin, op. cit.
17. For further details of Hathcock (and sniping in Vietnam) see Henderson, *Marine Sniper – 93 Confirmed Kills*, and Roberts and Sasser, *One Shot – One Kill*.
18. Roberts and Sasser, op. cit., p. 169.
19. Roberts and Sasser, op. cit., pp. 171–2.
20. Anon. Marine officer, quoted in *Firepower*, Issue 10, p. 219.
21. Ward, op. cit., pp. 38–43. Ward's experiences at Camp Pendleton cover the period February 1969.
22. See 3rd Marine Division Order 3590.3B: Scout-Sniper Platoons. This order provides details of the role, organization and tactical employment of snipers.
23. Ward, op. cit., pp. 59–60.
24. See 3rd Marine Division Order 3590.3B.
25. Reproduced in Chandler and Chandler, *Death From Afar*, p. 90.

26. Ward, op. cit., pp. 224–6.
27. 3rd Marine Division Order 3590.3B: Scout-Sniper Platoons, op. cit. p. 2.
28. US Army Concept Team in Vietnam: Sniper Operations and Equipment, p. 10.
29. US Army Combat Development Command: Sniper Programs, Trip Report, pp. 33–69.
30. Ewell and Hunt, *Sharpening the Combat Edge*, p. 121.
31. Ewell and Hunt, op. cit., p. 122.
32. Ewell and Hunt, op. cit., p. 123.
33. Hay, *Tactical and Material Innovations*, p. 75.
34. Hay, op. cit., p. 74.
35. Quoted in Johnson, 'Army and Marine Snipers . . .', in *Vietnam*, December 1989.
36. Ward, op. cit., p. 139.
37. Henderson, op. cit., pp. 1–2.

CHAPTER TEN
BRITISH SNIPING AFTER 1945

1. See Skennerton, *The British Sniper*, p. 145.
2. Anonymous interview with author. See also James D. Ladd, *Royal Marine Commando* (Hamlyn, 1982), p. 147.
3. Skennerton, op. cit., pp. 158–67. See also Laidler with Skennerton, *An Armourer's Perspective: .303 No. 4(T) Sniper Rifle*.
4. Almond and Savill, 'SAS joins hunt for "one-shot one-kill" IRA sniper squad', *The Daily Telegraph*, 31 July 1993.
5. Quoted in 'US Army Center for Lessons Learned, Bulletin 1–88, p. 3.
6. Ken Lukowiak, *A Soldier's Song* (Secker & Warburg, 1993), pp. 57–8.
7. See Peters, 'Handfeuerwaffen im Südatlantik', in *DWJ*, October 1982.
8. Quoted in 'US Army Center for Lesson Learned', Bulletin 1–88, p. 5.
9. Fraser, quoted in Middlebrook, *Operation Corporate*, p. 363.
10. Norris, 'The British Army's New Sniper Rifle' in *Armed Forces*, August 1986.

CHAPTER ELEVEN
THE BASICS OF SNIPER TRAINING

1. 'Infantry Training – Skill At Arms: Sniping', 1976. The information for this chapter has, for the most part, been gleaned from US and British sniper manuals and from conversations with sniper instructors from army and marine services.
2. 'Sniper Training and Employment', TC 23–14, June 1989, pp. 1–7.
3. Hathcock, quoted in Roberts, 'Master Sniper's One Shot Saves Lives', in *Soldier of Fortune*, May 1989.

4. Data from Plaster, *The Ultimate Sniper*, p. 293.

5. 'Sniper Training and Employment', TC 23–14, June 1989, chapter 5, p. 1.

CHAPTER TWELVE
THE SNIPER TRAINING COURSE

1. The section dealing with the Royal Marines' Badge Test is based on field interviews with RM instructing staff at Lympstone and the RM Sniper Badge Test pamphlet. See also McIntyre, 'The Sniper, Parts 1/2' in *Combat and Survival Magazine*, July 1992 and August 1992.

2. Maniprasad Rai, 'Sniper Training'.

3. Thompson, 'Scout-Sniper School', in *Leatherneck*, March 1984.

4. Walsh, quoted in Smith, 'Sniper School's Credo: "One Shot, One Kill"', in *The Washington Post*, n.d.

5. Daniel, 'USMC Scout Snipers: Issues and Answers', in *Marine Corps Gazette*, July 1990. See also Walsh, 'Reorganizing Scout Sniper Training' in *Marine Corps Gazette*, July 1990.

6. For a detailed study of Airborne sniper training, see Coleman, '"One Shot, One Kill" Army Sniper Training is Dead on Target', in *Soldier of Fortune*, December 1986.

7. Rozycki, 'US Army Sniper School', in *Infantry*, May–June 1989.

8. *SEAL Sniper Training Program*, p. 1.

CHAPTER THIRTEEN
SNIPING IN RECENT CONFLICTS

1. See Kokalis, 'Galil's New Sniper Rifle', in *Soldier of Fortune*, April 1987.

2. Interview by Dale A. Dye in 'Chuck Kramer: IDF's Master Sniper', in *Soldier of Fortune*, September 1985.

3. See USMC Development Center, untitled anonymous document, p. 2, held at USMC Museum Library, Washington; and Sasser and Roberts, *One Shot – One Kill*, pp. 248–50.

4. Roberts, 'American Snipers in Beirut', in *Soldier of Fortune*, August 1989.

5. 'US Army Center for Lessons Learned', Bulletin 1–88.

6. Chandler and Chandler, *Death From Afar*, p. 177.

7. See Kokalis, 'US Army Adopts Remington M24 SWS', in *Soldier of Fortune*, July 1988.

8. Feibel, 'Panama Kill', in *Soldier of Fortune*, May 1990.

9. See Plaster, *The Ultimate Sniper*, pp. 215–30, for further details of .50-calibre shooting, including the latest advances made by civilian shooters.

10. Chandler and Chandler, op. cit., p. 7.

11. Quoted in Chris Hedges, 'War is Vivid in Gun Sights of Sniper', *New York Times*, 3 February 1991.

12. Plaster, op. cit., p. 417.

13. Isby, *Weapons and Tactics of the Soviet Army*, pp. 409–10, 422–3; and Plaster, op. cit., pp. 390–2.
14. Quoted in Micheletti, 'Today's Snipers', in *Raids*, December 1993.
15. Hampstead, 'Third Time Lucky', in *Combat and Survival Magazine*, October 1992.

CHAPTER FOURTEEN
THE SNIPER TODAY

1. Anonymous, 'The Neglected Art of Sniping', in *Globe and Laurel*, May/June 1987.
2. Daniel, 'USMC Scout Snipers: Issues and Answers', in *Marine Corps Gazette*, July 1990.
3. For further details of night sniping, see Plaster, *The Ultimate Sniper*, pp. 405–15.
4. See 'Whispering Death', in *Combat and Survival Magazine*, May 1993.
5. Land, quoted in 'Sniper SNAFU', in *Soldier of Fortune*, March 1989.
6. For a more detailed discussion of the M118 round, see Kokalis, 'Sniper School', in *Soldier of Fortune*, September 1991; Daniel, op. cit.; Plaster, op. cit., pp. 112–13; USMC Development Center, untitled anonymous document, p. 2.
7. Figures from Plaster, op. cit., p. 130.
8. 'Focus on Sniper Rifles', in *Military Technology*, November 1986. This article provides an excellent summary of recent sniper rifle development and tactical application.

CHAPTER FIFTEEN
RIFLES

1. 'Focus on Sniper Rifles', in *Military Technology*, November 1986, p. 25.
2. Data comes from many sources. Apart from manufacturers' figures, the following were consulted: Hogg (ed), *Jane's Infantry Weapons 1990–91* (16th edition); Hogg, *Encyclopedia of Modern Small Arms*; Plaster, *The Ultimate Sniper*; Owen, *NATO Infantry and its Weapons*; Lonsdale, *Sniper/Counter-Sniper*; Isby, *Weapons and Tactics of the Soviet Army*; Baud, *Warsaw Pact Weapons Handbook*; Skennerton *The British Sniper*; Morrow, *The Confederate Whitworth Sharpshooters*; Senich, *The German Sniper 1914–45*; *Soldiers of Fortune* Magazine (Issues: May 1984, August 1986, September 1986, April 1987, July 1988, September 1989, August 1992); *War Machine* Partwork (Issues 65 and 86).

CHAPTER SIXTEEN
AMMUNITION

1. See Courtney-Green, *Ammunition for the Land Battle*, pp. 24–40; Goad and Halsey, *Ammunition*, pp. 27–30; Hogg, *The Illustrated Encyclopedia of Ammunition*; and Steadman, 'Trends in Small-arms Ammunition', in *Defence*, July 1985.

2. Armstrong, *Fieldcraft, Sniping and Intelligence*, p. 87.

3. Stevens, 'Safe Usage and Storage of Reloading Components' in *Target Gun*, February 1993; and Courtney-Green, op. cit., pp. 28–9.

BIBLIOGRAPHY

BOOKS

Armstrong, N. A. D., *Fieldcraft, Sniping and Intelligence* (Gale and Polden, 1942).

Ashworth, Tony, *Trench Warfare 1914–18: The Live and Let Live System* (Macmillan, 1980).

Baud, Jacques F., *Warsaw Pact Weapons Handbook* (Paladin Press, 1989).

Bolotin, D. N., *Fifty Years of Soviet Small Arms* (Department of the [US] Army, 1976).

Caputo, Philip, *A Rumour of War* (Arrow Books, 1978).

Cendrars, Blaise, *Lice* (Peter Owen, 1973).

Chandler, N. A., and Chandler, Roy, *Death From Afar: Marine Corps Sniping* (Iron Brigade Armory, Maryland, 1992).

Chesney, Lieutenant-Colonel C. H. R., *The Art of Camouflage* (Robert Hale, 1941).

Courtney-Green, P. R., *Ammunition for the Land Battle* (Brassey's, 1991).

Crum, F. M., *Scouts and Snipers in Trench Warfare* (privately printed, 1916).

Doubler, Michael D., *Busting the Bocage: American Combined Operations in France, 6 June–31 July 1944* (Combat Studies Command, US Army Command, 1988).

Dunn, J. C., *The War the Infantry Knew 1914–19* (Cardinal, 1987).

Edmonds, Charles (Carrington, C. E.), *A Subaltern's War* (Anthony Mott, 1984).

Ellis, John, *The Sharp End of War: The Fighting Man in World War II* (David & Charles, 1980).

Ewell, Julian J., and Hunt, Ira A., *Sharpening the Combat Edge: The Use of Analysis to Reinforce Military Judgement* (Vietnam Studies, Department of the Army, 1974).

Fatteh, Abdullah, *Medicolegal Investigation of Gunshot Wounds* (J. B. Lippincott, 1976).

Goad, K. J. W., and Halsey, D. H. J., *Ammunition* (Brassey's, 1982).

Graves, Robert, *Goodbye to All That* (Penguin, 1981).

Greener, W. W., *Sharpshooting for Sport and War* (R. A. Everett, 1900).

Hartcup, Guy, *Camouflage: A History of Concealment and Deception in War* (David & Charles, 1979).

Hay, John H., Jnr, *Tactical and Material Innovations* (Vietnam Studies, Department of the Army, 1974).

Haythornthwaite, Philip J., *The Napoleonic Source Book* (Arms and Armour Press, 1990).

Henderson, Charles, *Marine Sniper: 93 Confirmed Kills* (Berkeley, 1988).

277

Hermes, Walter G., *Truce Tent and Fighting Front: US Army in the Korean War* (Office of the Chief of Staff of Military History, US Army, 1966).

Herr, Michael, *Despatches* (Picador, 1979).

Hesketh-Prichard, H., *Sniping In France: With Notes on the Scientific Training of Scouts, Observers and Snipers* (Hutchinson, n.d.).

Hogg, Ian V., *Encyclopedia of Modern Small Arms* (Hamlyn/Bison, 1983).

Hogg, Ian V. (ed.), *Jane's Infantry Weapons 1990–91* (Jane's Publishing, 1991).

Hogue, Oliver, *Trooper Bluegum in the Dardanelles* (Melrose, 1916).

Idriess, Ion L., *The Desert Column* (Angus and Robertson, 1932).

Isby, David C., *Weapons and Tactics of the Soviet Army* (Jane's, 1988).

Jünger, Ernst, *Copse 125* (Chatto and Windus, 1930).

Laidler, Peter, with Skennerton, Ian, *An Armourer's Perspective: .303 No. 4(T) Sniper Rifle and the Holland and Holland Connection* (Skennerton, 1993).

Lewis, Berkeley R., *Small Arms and Ammunition in the United States Service* (Smithsonian Institute, 1956).

Lonsdale, Mark V., *Sniper/Counter-Sniper* (STTU, 1987).

—— *Advanced Weapon Training* (STTU, 1988).

Lumsden, Malvern (ed.), *Anti-Personal Weapons* (SIPRI/Taylor and Francis, 1978).

Marcot, Roy M., *Civil War Chief of Sharpshooters: Hiram Berdan – Military Commander and Firearms Inventor* (Northwood Heritage Press, 1987).

McBride, Herbert W., *A Rifleman Went to War* (Lancer Militaria, 1987).

Melville, Michael Leslie, *The Story of the Lovat Scouts* (The Saint Andrew Press, 1981).

Middlebrook, Martin, *Operation Corporate: The Falklands War 1982* (Viking, 1985).

Moore, Warren, *Weapons of the American Revolution* (Promontory Press, 1967).

Morrow, John Anderson, *The Confederate Whitworth Sharpshooters* (privately published, 1989).

Neumann, George C., *The History of the Weapons of the American Revolution* (Bonanza Books, 1967).

Owen, J. I. H., *Nato Infantry and its Weapons* (Brassey's Publishing, 1976).

Owen-Smith, M. S., *High-Velocity Missile Wounds* (Edward Arnold, 1981).

Peterson, Harold L., *The Book of the Gun* (Hamlyn, 1968).

Page, Tim, and Pimlott, John, *NAM: The Vietnam Experience 1965–75* (Orbis/Hamlyn, 1988).

Plaster, John L., *The Ultimate Sniper* (Paladin Press, 1993).

Reitz, Deneys, *Commando: A Boer Journal of the Boer War* (Faber and Faber, 1929).

Richards, Frank, *Old Soldiers Never Die* (Anthony Mott, 1983).

Roberts, Craig, and Sasser, Charles W., *The Walking Dead: A Marine's Story of Vietnam* (Grafton, 1990).

Sasser, Charles W., and Roberts, Craig, *One Shot – One Kill* (Pocket Books, 1990),

Sawyer, Charles Winthrop, *Firearms in American History* (privately published, 1910).

Senich, Peter R., *Limited War Sniping* (Paladin Press, 1977).

—— *The Pictorial History of US Sniping* (Paladin Press, 1980).

—— *The German Sniper 1914–45* (Paladin Press, 1982).

Shores, C., *With British Snipers to the Reich* (Lancer Militaria, 1988).

Shulimson, Jack, and Johnson, Charles M., *US Marines in Vietnam 1965* (History and Museums Division Headquarters, USMC, 1978).

Skennerton, Ian, *The British Sniper: British & Commonwealth Sniping and Equipments 1915–1983* (Skennerton/Arms & Armour Press, 1984).

Sword, Wiley, *Sharpshooter: Hiram Berdan, his famous Sharpshooters and their Sharps Rifles* (Andrew Mowbray Inc., 1988).

Tantum, William H., *Sniper Rifles of the Two World Wars* (Ottawa Museum Restoration Service, 1967).

Thomas, John W., Jnr, *Fix Bayonets!* (Scribners, 1926).

Truby, J. David, *Silencers, Snipers and Assassins* (Paladin, 1972).

Ward, Joseph T., *Dear Mom: A Sniper's Vietnam* (Ivy Books, 1991).

Westerfield, Hargis, *The Jungleers: A History of the 41st Infantry Division* (41st Association, 1980).

JOURNALS, MAGAZINES, NEWSPAPERS AND OTHER PERIODICALS

Allen, Henry, 'The Snipers in the Desert, Eyes on the Target' in *The Washington Post*, 28 December 1990.

Almond, Peter, and Savill, Richard, 'SAS Joins Hunt for "One-Shot One-Kill" IRA Sniper Squad' in *The Daily Telegraph*, 31 July 1993.

Amos, Albert A., Jnr, '.50 Caliber Sniper' in *Infantry*, September–October 1970.

Bull, Stephen, 'British Army Snipers 1914–18' in *Military Illustrated*, September 1982.

Burks, Arthur J., 'The Sniper' in *Leatherneck*, August 1926.

Bruce, Robert, 'Test Firing the K43 Rifle' in *Combat and Survival Magazine*, February 1991.

Childs, Jack, 'VC Defector Unveils Elite Sniper Company' in *Sea Tiger*, 19 July 1966.

Childs, Jack, and Martin, Bruce, 'Sniper . . .' in *Leatherneck*, January 1967.

Clapham, Richard, 'Sniping and Observing in Wartime' in *Cavalry Journal*, vol. XXX, 1940.

Coleman, John, 'One Shot, One Kill: Army Sniper School is Dead on Target' in *Soldier of Fortune*, December 1986.

Collins, Edward J., 'Wanted: A Killer/Pro & Con' in *Infantry*, January–March 1959.

Crupi, A. J., and Odom, William F., 'A Warning' in *Infantry School Quarterly*, October 1951.

Daniel, Edward D., 'USMC Scout Snipers: Issues and Answers' in *Marine Corps Gazette*, July 1990.

Dye, Dale A., 'Chuck Kramer: IDF's Master Sniper' in *Soldier of Fortune*, September 1985.

Enoch, Leslie B., 'Sniper – Sayonara/Pro & Con' in *Infantry*, January–March 1959.

Feibel, Bobby, 'Panama Kill: Vietnam Sniper Runs Count to 39' in *Soldier of Fortune*, May 1990.

'Focus on Sniper Rifles' in *Military Technology*, November 1986.

Grigg, Weldon M., 'One Shot – One Kill' in *Infantry*, November–December 1979.

Hampstead, Skippy, 'Third Time Lucky . . .' in *Combat and Survival Magazine*, October 1992.

Hargreaves, Reginald, 'The Lonely Art' in *Marine Corps Gazette*, December 1954.

Hedges, Chris, 'War is Vivid in Gun Sights of the Sniper' in *The New York Times*, 3 February 1991.

Hicks, Norman W., 'Team Shots Can Kill' in *Marine Corps Gazette*, December 1963.

Hofues, John L., 'Modernize the Sniper Rifle' in *Army*, June 1957.

Hogg, Ian V., 'Aiming to Kill' in *War in Peace* (Orbis Publishing), No. 17.

'How the Japs Train Snipers' in *Military Review*, October 1945.

Johnson, David A., 'Army and Marine Snipers . . .' in *Vietnam*, December 1989.

Karwan, Chuck, 'Sniper SNAFU: US Army's M24 SWS Defeated by Design' in *Soldier of Fortune*, March 1989.

Kokalis, Peter G., 'ComBloc Sniper Rifles' in *Soldier of Fortune*, August 1986.

—— 'HK BASR' in *Soldier of Fortune*, September 1986.

—— 'Galil's New Sniper Rifle' in *Soldier of Fortune*, April 1987.

—— 'Sniper Systems . . . M24 SWS [and] Parker/Hale M85' in *Soldier of Fortune*, July 1988.

—— 'To Kill or Incapacitate: A Thinking Man's Guide to Wound Ballistics' in *Soldier of Fortune*, Fighting Firearms Special.

—— 'Killing Effect II: Rifleman's Guide to Wound Ballistics' in *Soldier of Fortune*, January 1989.

—— 'Take-Down Sniper System: H–S Precision's New Convertible Match-Accurate Magnum' in *Soldier of Fortune*, September 1989.

—— 'Sniper School' in *Soldier of Fortune*, September 1991.

—— 'Death in the Tall Grass: Father of M16 Forges Marine-Spec Sniper System' in *Soldier of Fortune*, August 1992.

Kriventsov, M., 'Soviet Snipers' in *Infantry Journal*, October 1942.

Lewis, Jack, 'One Course Open: Slay' in *Gun World* [date unknown].

Linwood, R. J., 'The Sniper – Part 1' [journal and date unknown].

'Marine Sniper Picks Off VC' in *The Observer* (Saigon), 16 April 1966.

Martin, Bruce, '4th Marines Snipe at 1000 Yards Range' in *Sea Tiger*, 26 July 1966.

Martin, Glen E., 'They Call Their Shots' in *Marine Corps Gazette*, April 1953.

McIntyre, Mick, 'The Sniper Part 1' in *Combat and Survival Magazine*, July 1992.

—— 'The Sniper Part 2' in *Combat and Survival Magazine*, August 1992.

McKnight, John T., 'Snipers' in *Infantry*, September–October 1969.

Micheletti, Eric, and Maes, Frederic, 'The New Sniping Threat' in *Raids*, December 1993.

Murphy, Jack, 'Snipers Learn a Deadly Art' in *Washington Morning Star*, 8 July 1985.

Norris, John, 'The British Army's New Sniper Rifle' in *Armed Forces*, August 1986.

Observer, 'Snipers Are Overrated' in *Marine Corps Gazette*, January 1945.

Odom, William F., 'The Case of the US Sniper' in *Infantry School Quarterly*, April 1954.

Ogle, Clarence O., 'Let's Get the Most from our Shooters' in *Army*, February 1957.

Peters, John, 'Handfeuerwaffen im Südatlantik' in *Deutsches Waffen Journal*, October 1982.

Rai, Maniprasad, 'Sniper Training' [journal and date unknown].

'Rifles of the Great War' in *War Machine* (Orbis/Aerospace Publishing), No. 86.

'Rifles of World War II' in *War Machine* (Orbis/Aerospace Publishing), No. 65.

Roberts, Craig, 'Master Sniper's One Shot Saves Lives' in *Soldier of Fortune*, May 1989.

—— 'American Snipers in Beirut' in *Soldier of Fortune*, August 1989.

—— 'At Stalingrad . . . Sniper versus Sniper' in *World War II*, September 1989.

Rogers, Glenn F., Jnr, and Hackney, Michael S., 'MILES Sniper Training' in *Infantry*, March–April 1983.

Rozycki, Mark L., 'US Army Sniper School' in *Infantry*, May–June 1989.

Satterwhite, John, 'Terrorists Beware: H&K Takes Aim with New PSG–1' in *Soldier of Fortune*, May 1984.

Silkett, Wayne A., 'Urban Snipers' in *Infantry*, September–October 1982.

Simpson, Ross, 'Cross-Hairs on Baghdad: USMC Scout Snipers Wait to Reach and Touch Someone' in *Soldier of Fortune*, March 1991.

Sines, Kenneth A., 'What's in the Future for Snipers?' in *Infantry*, May–June 1972.

Smith, Philip, 'Sniper's Credo: "One Shot, One Kill"' in *The Washington Post* [date unknown].

'The Sniper' in *Military Review*, September 1944.

Steadman, Nick, 'Trends in Small-arms Ammunition' in *Defence*, July 1985.

Stevens, Clive, 'Safe Usage and Storage of Reloading Components' in *Target Gun*, February 1993.

Taylor, Chuck, 'The Sniper and his Tools' [magazine and date unknown].

Thompson, P. L., 'Scout-Sniper School' in *Leatherneck*, March 1984.

Tolbert, Frank X., 'Deadly Teams Emerge From This Academy' in *Leatherneck*, October 1943.

'US Army Center For Lessons Learned', *Bulletin No. 1–88*, April 1988.

Walsh, Steven L., 'Reorganizing Scout Sniper Training' in *Marine Corps Gazette*, July 1990.

'Whispering Death' in *Combat and Survival Magazine*, May 1993.

Wright, D. L., 'Training the Scout Sniper' in *Marine Corps Gazette'*, October 1985.

MANUALS, PAMPHLETS AND OTHER OFFICIAL DOCUMENTS

Canadian Army, sniper training manual (CFP 309 [5]), chapters 4 (Field Training) and 5 (Sniper Employment).

Division Order 3590.3B: Scout-Sniper Platoons, HQ 3rd Marine Division, 9 June 1968 (USMC Museum, Washington DC).

German Army Handbook, April 1918 (Arms & Armour reprint, 1977).

Infantry Training Volume 1: Skill at Arms, Pamphlet No. 4 Sniping (Crown Copyright, 1976 and 1990) [UK].

Military Training Pamphlet No. 44 – 1940: Notes on the Training of Snipers [UK].

SEAL Sniper Training Program (Paladin Press reprint).

Scout Snipers in Current Operations, HQ Marine Corps Routing Sheet/ A03M–jaj, 26 January 1966 (USMC Museum, Washington DC).

Siege Operations Committee: Report on the Dungeness Experiment of 1879, Appendix A (Pattern Room, Nottingham).

Sniper Badge Test notes [RM, UK].

Sniper Weapon System Requirements to Annex J to the Final Report on Infantry Organization and Weapon Systems, 1973–77 (USMC Museum, Washington DC).

Test Report of Marine Corps Equipment Board: Sniper Rifles, Telescopes and Mounts, 31 August 1951 (USMC Museum, Washington DC).

US Army Combat Development Command: Trip Report, Sniper Programs, 28 April 1969 (US Military History Institute, Carlisle, Pa).

US Army Concept Team in Vietnam: Sniper Operations and Equipment, Final Report, 23 February 1968 (US Military History Institute, Carlisle, Pa.).

US Army Counter-Sniper Guide.

US Army Ground Forces, Observer Board, Southwest Pacific Ocean Area: Training and Use of Snipers, 5 January 1945 (US Army Military History Institute, Carlisle, Pa.).

US Army Sniper Training and Employment, TC 23–14.

USMC Development Center: untitled and undated document referring to USMC sniper rifle development in 1970s (USMC Museum, Washington DC).

USMC *Sniping* manual (Desert Publications reprint).

US Military Assistance Command, Vietnam, Combined Intelligence Center Vietnam: VC/NVA Employment of Snipers, 6 January 1967.

OTHER UNPUBLISHED SOURCES

Baker, Thomas: audio interview, Department of Sound Records, Imperial War Museum.

Carson Catron, W.: 'For King and Country', Department of Documents, Imperial War Museum.

Clarke, B. A.: 'Sniper on the Western Front', Department of Documents, Imperial War Museum.

Ewell, Julian J.: Senior Officer Debriefing Program, US Army Military History Institute, Carlisle, Pa.

Glover, F. P. J.: ms diary, Department of Documents, Imperial War Museum.

Hall, Percy Raymond: 'Recollections 1915–19', Department of Documents, Imperial War Museum.

Hills, Mr: audio transcript from 'The Great War' (BBC), Department of Sound Records, Imperial War Museum.

Jalland, William: audio interview, Department of Sound Records, Imperial War Museum.

McCowan, A.: ms diary, Department of Documents, Imperial War Museum.

Rabbets, Edgar: audio transcript interview, Department of Sound Records, Imperial War Museum.

Ricketts, V. G.: ms memoirs, Department of Documents, Imperial War Museum.

Spearman, William: audio transcript interview, Department of Sound Records, Imperial War Museum.

Stainton, Harold: 'A Personal Narrative of the War', Department of Documents, Imperial War Museum.

Van Orden, George O., and Lloyd, Calvin A.: 'Equipment for the American Sniper', USMC Museum, Washington DC.

INDEX